Silicon Materials

Edited by Beddiaf Zaidi and Slimen Belghit

Published in London, United Kingdom

IntechOpen

Supporting open minds since 2005

Silicon Materials
http://dx.doi.org/10.5772/intechopen.81584
Edited by Beddiaf Zaidi and Slimen Belghit

Contributors
Siti Salina Sarkawi, Zhi Peng Ling, Zheng Xin, Rolf Stangl, Adolfo Romero, Denis Cabrera, Lucero Rosales-Marines, Anilu Rubio-Ríos, Lorena Farías-Cepeda, Jose Edgar Alfonso Orjuela, Henry Samir Vanegas, Jairo Olaya, Luong Thi Kim Phuong, Mujtaba Ikram, Muhammad Umer Farooq, Beddiaf Zaidi, Alexander Chaika, Victor Aristov, Olga Molodtsova

First published in London, United Kingdom, 2019 by IntechOpen
IntechOpen is the global imprint of INTECHOPEN LIMITED, registered in England and Wales, registration number: 11086078, The Shard, 25th floor, 32 London Bridge Street
London, SE19SG - United Kingdom
Printed in Croatia

British Library Cataloguing-in-Publication Data
A catalogue record for this book is available from the British Library

Additional hard and PDF copies can be obtained from orders@intechopen.com

Silicon Materials
Edited by Beddiaf Zaidi and Slimen Belghit
p. cm.
Print ISBN 978-1-78984-658-4
Online ISBN 978-1-78984-659-1
eBook (PDF) ISBN 978-1-78985-570-8

For EU product safety concerns:
IN TECH d.o.o., Prolaz Marije Krucifikse Kozulić 3, 51000 Rijeka, Croatia, info@intechopen.com or visit our website at intechopen.com.

Meet the editors

Dr. Beddiaf Zaidi is an assistant professor in the Department of Physics at the University of Batna 1, Algeria. He obtained a PhD in Physics from the University of Annaba, Algeria, in 2014. Dr. Zaidi has published a number of research papers in reputed journals and has written a book on solar cells. He is also a peer reviewer for journal articles. He previously served as an editor-in-chief of the *International Journal of Materials Science and Applications* (IJMSA). He has participated in many international conferences serving as a referee and a program committee member. He is on the editorial boards of numerous journals and is a lead guest editor for many special issues.

Belghit Slimen obtained a PhD in Radiation Physics from the University of Batna, Algeria. He is a member of the interaction laboratory at the University of Batna 1, Batna, Algeria. He has authored several scientific articles and journal papers and has participated in many international conferences.

Contents

Preface

Silicon is a chemical element with the symbol Si and atomic number 14. It is a hard and brittle crystalline solid with a blue-grey metallic luster, and it is a tetravalent metalloid and semiconductor.

Most silicon is used commercially without being separated, and often with little processing of the natural minerals. It has many applications in the manufacturing technology of microelectronic components, integrated circuits, and photovoltaic generators. Circuit complexity and higher degrees of integration of components require constant improvement and control of the properties of silicon. Elemental silicon also has a large impact on the modern world economy.

We wish to thank everyone who showed us patience and understanding and offered us help during the intensive phase of creating this book.

Dr. Beddiaf Zaidi
Department of Physics, Faculty of Material Sciences,
University of Batna 1, Batna, Algeria

Dr. Slimen Belghit
University of Batna 1, Batna, Algeria

Chapter 1

Introductory Chapter: Silicon

Beddiaf Zaidi and Slimen Belghit

1. Introduction

Silicon (Si) is a member of Group 14 (IV_A) in the periodic table of elements. Si is also part of the carbon family. This family elements include C, Ge, Sn, and Pb. Silicon is a metalloid, one of only a very few elements that have properties of both metals and non-metals. Si is the second most abundant element in the Earth's crust, apart from oxygen.

2. Energy bands

Silicon is a semiconductor whose number of free electrons is less than conductor but more than that of an insulator. Two kinds of energy band which are conduction and valence. Series of energy levels having valence electrons forms valance band in solid. At 0°K, the energy levels of valence band are filled with electrons. This band contains maximum of energy when the electrons are in valence band. Conduction band is the higher energy level band, which is the minimum of energy. Conduction band is partially filled by the electrons, which are known as the free electrons as they can move anywhere in solid. These electrons are responsible for current flowing. There is a gap of energy. The difference between the conduction band and valence band is called energy gap. For semiconductors, the gap is neither large nor the bands get overlapped (**Figure 1**).

3. Basic material

Apart from the oxygen, silicon is most commonly occurring element on the Earth. Silica is the dioxide from silicon and occurs mostly as quartz. Its synthesis has been familiar for many decades. It is extracted from (mainly) quartzite reduction with carbon in an arc furnace process [1]. The pulverized quartz and carbon are put in a graphite crucible. An arc causes them to melt at approximately 1800°C. Then, the reduction process takes place according to the formula:

$$SiO_2 + 2C \rightarrow Si + 2CO_{(g)} \quad (1)$$

The liquid collected at the bottom of crucible can then be drawn off. Its purity can be approximately 97.9%. This is called metallurgic grade silicon (MG-Si). However, for silicon to be used in the semiconductor industry, the impurities must be removed almost completely by further processes. For such a high purity grade, multistage processes must be implemented.

IntechOpen

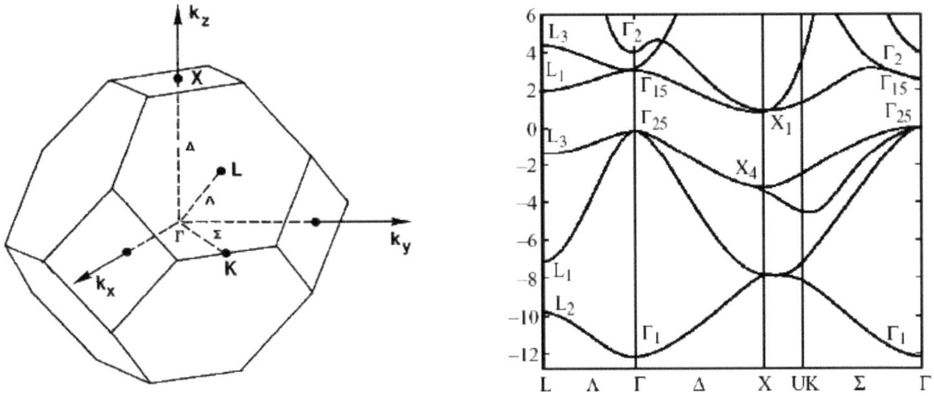

Figure 1.
E(k) for the conduction and valence bands of silicon.

4. Silicon thin film

Silicon thin-film cells are mainly deposited by chemical vapor deposition (typically plasma-enhanced, PE-CVD) from silane gas and hydrogen gas. Depending on the deposition parameters, these silicon thin films can be based on one or a combination of these materials [2–5]:

1. Amorphous silicon (a-Si or a-Si:H) or polymorphous silicon

2. Microcrystalline silicon

3. Polycrystalline silicon (poly-Si).

These silicon thin film materials can be characterized by their grain sizes ranging from none (amorphous) to large silicon (~100 μm) for polysilicon. The crystalline silicon thin films present dangling bonds, which result in deep defects (energy levels in band gap) as well as deformation of the conduction and valence bands. The solar cells made from thin films tend to have lower energy conversion efficiency than silicon bulk but are also less expensive to produce (**Table 1**).

Type of silicon	Abbreviation	Crystal size	Deposition method
Single-crystal silicon	Sc-Si	>10 cm	Czochralski, Float zone
Multicrystalline silicon	Mc-Si	1 mm–10 cm	Cast, sheet, ribbon
Polycrystalline silicon	Poly-Si	1 μm–1 mm	Chemical-vapor deposition (at high temperature ≥ 1000°C)
Microcrystalline silicon	μc-Si	10 nm–1 μm	Ex.: Plasma deposition (at low temperature < 600°C)
Nanocrystalline silicon	nc-Si	1–10 nm	

Table 1.
Grain size range depending on the type of the silicon [6–8].

5. Application

Silicon materials are used in components of electronic devices. It is also used to make solar cells [9–13] and parts for computer circuits [14]. Solar cell is a device that converts sun light into electrical energy [15–23]. A rectifier is an electrical device that converts alternating current to direct current. The most important silicon alloys are those made with Fe, Al, and Cu. When silicon is produced, in fact, scrap iron and metal are sometimes added to the furnace [24, 25].

Author details

Beddiaf Zaidi* and Slimen Belghit
Department of Physics, Faculty of Material Sciences, University of Batna 1, Batna, Algeria

*Address all correspondence to: zbeddiaf@gmail.com

References

[1] Deal BE, Grove AS. General relation for the oxidation of silicon. Journal of Applied Physics. 1965;**36**:3770

[2] Zaidi B, Hadjoudja B, Chouial B, Kamli K, Chibani A, Shekhar C. Impact of hydrogen passivation on electrical properties of polysilicon thin films. Silicon. 2018;**10**:2161

[3] Zaidi B, Shekhar C, Hadjoudja B, Chouial B, Li R, Madhava Rao MV, et al. Dopant segregation and heat treatment effects on the electrical properties of polycrystalline silicon thin films. Silicon. 2016;**8**:513

[4] Ghitani HE, Pasquinelli M, Martinuzzi S. Influence of dislocations on photovoltaic properties of multicrystalline silicon solar cells. Journal de Physique III. 1993;**3**(19):41

[5] Zaidi B, Shekhar C, Hadjoudja B, Chouial B, Chibani A, Li R, et al. Optimum parameters for obtaining polycrystalline silicon for photovoltaic application. American Journal of Nanosciences. 2015;**1**(1)

[6] Jellison GE, Modine FA, White CW, Wood RF, Young RT. Optical properties of heavily doped silicon between 1.5 and 4.1 eV. Physical Review Letters. 1981;**46**:1414

[7] Zaidi B, Hadjoudja B, Chouial B, Gagui S, Felfli H, Magramene A, et al. Effect of secondary annealing on electrical properties of polysilicon thin films. Silicon. 2015;**7**:293

[8] Zaidi B, Hadjoudja B, Chouial B, Gagui S, Felfli H, Chibani A. Hydrogenation effect on electrical behavior of polysilicon thin films. Silicon. 2015;**7**:275

[9] Moller HJ. Semiconductors for Solar Cells. Norwood, MA: Artech House. Inc.; 1993

[10] Zaidi B. Introductory chapter: Introduction to photovoltaic effect. In: Solar Panels and Photovoltaic Materials. London, UK: InTech Open; 2018. pp. 1-8

[11] Zaidi B, Hadjoudja B, Felfli H, Chouial B, Chibani A. Effet des Traitements Thermiques sur le Comportement Électrique des Couches de Silicium Polycristallin pour des Applications Photovoltaïques. Revue de Métallurgie. 2011;**108**:443

[12] Goetzberger A, Knobloch J, Voss B. Crystalline Silicon Solar Cells. England: Wiley; 1998

[13] Zaidi B, Hadjoudja B, Felfli H, Chibani A. Influence of doping and heat treatments on carriers mobility in polycrystalline silicon thin films for photovoltaic application. Turkish Journal of Physics. 2011;**35**:185

[14] Mathieu H. physique des semi-conducteurs et des composants électroniques. Paris: Masson; 1987

[15] Zaidi B, Saouane I, Shekhar C. Electrical energy generated by amorphous silicon solar panels. Silicon. 2018;**10**:975

[16] Zaidi B, Saouane I, Shekhar C. Simulation of single-diode equivalent model of polycrystalline silicon solar cells. International Journal of Materials Science and Applications. 2018;**7**(8)

[17] Conibeer G. Third-generation photovoltaics. Materials Today. 2007;**10**:42

[18] Zaidi B, Belghit S, Shekhar C, Mekhalfa M, Hadjoudja B, Chouial B. Electrical performance of $CuInSe_2$ solar panels using ant colony optimization algorithm. Journal of Nano- and Electronic Physics. 2018;**10**:05044

[19] Haddad A, Inokuma T, Kurata Y, Hasegawa S. Characterization of structure and role of different textures in polycrystalline Si films. Journal of Non-Crystalline Solids. 2005;**351**:2107

[20] Zaidi B, Saouane I, Madhava Rao MV, Li R, Hadjoudja B, Gagui S, et al. Matlab/simulink based simulation of monocrystalline silicon solar cells. International Journal of Materials Science and Applications. 2016;**5**(11)

[21] Jayawardena K, Rozanski LJ, Mills CA, Beliatis MJ, Nismy NA, Silva S. Inorganics-in-organics': Recent developments and outlook for 4G polymer solar cells. Nanoscale. 2013;**5**:8411

[22] Zaidi B, Belghit S, Shekhar C, Houaidji N, Hadjoudja B, Chouial B. Impact of anti-reflective coating on the characteristics of a-Si: H solar cells. Nanosistemi, Nanomateriali, Nanotehnologii. 2018;**16**:713

[23] Zaidi B, Hadjoudja B, Belghit S, Chouial B, Mekhalfa M. Annealing effect on grain boundary width of polycrystalline silicon for photovoltaic application. Revue des Energies Renouvelables. 2018;**21**:397

[24] Sigfússon T, Helgason Ö. Rates of transformations in the ferrosilicon system. Hyperfine Interactions. 1990;**54**:867

[25] Waanders FB, Mans A. Characterisation of ferrosilicon dense medium separation material. Hyperfine Interactions. 2003;**148**:325

Chapter 2

The Silicon on the Catalysis: Hydrodesulfurization of Petroleum Fractions

Denis A. Cabrera-Munguia, Lucero Rosales-Marines, Anilu Rubio-Ríos, Lorena Farías-Cepeda and Adolfo Romero-Galarza

Abstract

The use of silicon in hydrodesulfurization (HDS) catalysts has had an impact on some of the fundamental parameters for the performance of the HDS catalyst, such as dispersion of the active phase, the extent of sulfidation, or the level of promotion. Mesoporous silicates such as MCM-41, SBA-15, the modified surface of γ-alumina with silica, or mixed Al_2O_3-SiO_2 support have showed to affect the metal-support interaction, favoring the so-called type II Co(Ni)–Mo(W)–S structures. The aim of this chapter is to elaborate an analysis on the recent advances in the synthesis design of silicon catalyst support for the hydrodesulfurization of petroleum fractions.

Keywords: type I and type II structures, HDS, metal-support interaction, sulfidation, mesoporous silicates, silica

1. Introduction

The air pollution is one of the main problems that the governments around the world have been concerned to mitigate in recent decades. The harmful gases found in the atmosphere are the product of the generation of energy through the combustion of hydrocarbons [1–3]. However, the hydrocarbons will continue to be used as the main source of energy in the next decades [4]; thus, it represents an environmental problem. The main pollutants generated by the combustion of fuels are SOx, CO, NOx, and traces of some heavy elements. The SOx is considered particularly dangerous since SO_2 can be oxidized to SO_3 by several routes, depending on the particular conditions of the atmosphere. Once the SO3 is formed, it is diluted in the water droplets that are present in the atmosphere, therefore generating the sulfuric acid (H_2SO_4), leading to the acid rain. Moreover, it produces a direct environmental damage to humanity, such as eye irritation and constriction of the respiratory tract, causing harm to the entire population, but especially to asthmatics and other sensitive people. It also contributes to vegetation damage, causing discoloration and lesions on the leaves. In addition, sulfur dioxide has been associated with steel corrosion, deterioration of concrete structures, paper, leather, historical monuments, and certain textiles. Based on the described problems, countries have opted to make stringent environmental regulations to reduce the levels of air pollutant emissions. Worldwide legislations have been issued to reduce the amount of sulfur contents in the transport fuels close to zero sulfur ppm (ultralow sulfur fuels) [5]. This represents a challenge due to the declining trend of light oil supplies [6], leading to process heavy crudes with higher concentrations of sulfur, nitrogen, and metals. In this context, the hydrodesulphurization (HDS) is considered the most effective used process to produce ultralow sulfur transport fuels, for which the design and preparation of catalysts of high HDS performance are believed to be central factors [7, 8]. In the last decade, traditional HDS catalysts have usually been based on Mo or W sulfides promoted by Co or Ni supported on γ-alumina [9, 10]. Nonetheless, due to the need to process increasingly heavy crudes and in consequence with higher concentrations of sulfur as mention before, it is becoming more difficult to produce ultralow-sulfur transport fuels using traditional HDS catalysts [11, 12]. Therefore, it is necessary to design novel catalysts with higher performance in the HDS reaction. In order to achieve it, the HDS catalysts require (1) complete sulfidation of the molybdenum or tungsten and Co(Ni) precursor oxides phases, (2) high extent of promotion, and (3) high dispersion of the Co(Ni)Mo(W)S active phase. It is well known that the strength of interaction between the support and the Co(Ni)Mo(W)S active phase has an important effect on the above three parameters. Alumina interacts strongly with the Co, Ni, Mo, and W oxide-supported phases; therefore, new support materials with weaker metal-support interaction must be identified. This chapter will explain which factors can lower the reactivity for HDS in some sulfur compounds such as 4,6-DMDBT, as well as the models proposed to explain the nature of the active sites, and briefly summarize recent advances in the use the silicon in catalyst support with the aim of understanding the difference in HDS performance compared with the traditional catalyst supported on γ-alumina from the point of view of the strength of the interaction between the support and the active phase.

2. Hydrodesulfurization (HDS)

Hydrodesulfurization is one of the most important processes among the oil refining industry, whose purpose is to reduce pollutants in the fraction of the petroleum distillates. The operation conditions of HDS reaction are in a pressure range between 60 and 200 atm and temperatures higher than 280°C. On the other hand, the sulfur vacancies, called coordinatively unsaturated sites (CUS), present at the edge of the $Mo(W)S_2$ crystals are the active sites [13, 14].

Crude oil contains a complex mixture of sulfur compounds, which in turn have a different reactivity. It is known that for each type of fuel, the sulfur-containing molecules are different and the degree of reactivity of these sulfur compounds in HDS depends on its structure. The degree of reactivity for the removal of sulfur can vary in several magnitudes. Generally, acyclic sulfides such as sulfides, disulfides, and thiols are highly reactive in HDS compared to thiophene. The reactivities of sulfur compounds with 1–3 rings decrease in the following order thiophenes > benzothiophenes > dibenzothiophenes [14]. Similarly, the reactivity of alkyl-substituted compounds such as 4-methyldibenzothiophene and 4,6-dimethyldibenzothiophene (4,6-DMDBT) is much less reactive than other compounds. **Figure 1** qualitatively represents the relationship between the type and size of sulfur compound in different fractions of distillate and their relative reactivity. In the range of diesel, molecules such as dibenzothiophenes and alkyl-dibenzothiophenes predominate, which present low reactivity for HDS.

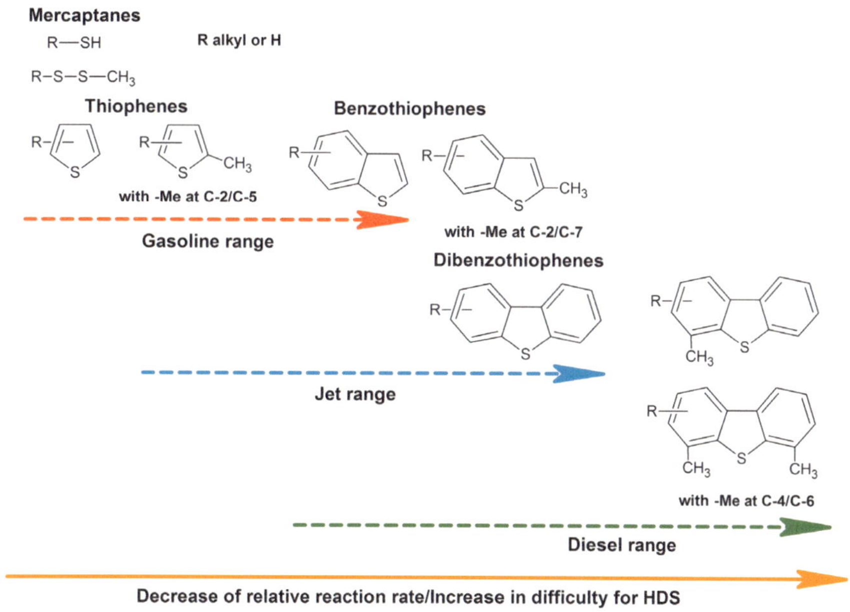

Figure 1.
Relative reaction rate of organic sulfur compounds.

2.1 Reaction routes

The activity study of the catalysts for deep desulfurization has focused mainly on the most refractory molecules as can be observed in the literature [14, 15]. These molecules such as 4,6-dimethyldibenzothiophene (4,6-DDMBT), 4-methyldibenzothiophene (4-MDBT), and dibenzothiophene (DBT) have different reaction routes [16, 17], considering mainly the direct desulfurization route (DDS) and the hydrogenation route (HYD).

Desulfurization route (DDS). In the case of 4,6-DMBT, the DDS route is one in which the sulfur atom is removed from the structure and replaced by hydrogen, without carrying out the hydrogenation of any of the carbon-carbon double bonds present in the molecule. Some authors [17, 18] proposed that once it is adsorbed, the molecule on the active site, the DDS route begins with hydrogenation of one of the double bonds adjacent to the sulfur atom, to obtain a dihydrogenated product and then, the opening of the C—S bond through a process of elimination. **Figure 2** shows the mechanism for the C—S cleaving. According to the studies of the structure reactivity, the activity for the DDS reactions is attributed to sites located at the edges of the MoS_2 crystal [19], and it is suggested that the sites are sulfur anionic vacancies, called coordinatively unsaturated sites (CUS), which are created and regenerated during the reaction in the presence of hydrogen.

Hydrogenation route (HYD). In the HDS, the HYD route involves the hydrogenation of one of the aromatic rings, prior to the cleaving of the C—S bond: assuming that the cleavage of the C—S bond occurs through the β-elimination process, several explanations for the low reactivity of 4,6-DMDBT have been proposed and discussed by Bataille et al. [17] such as: (a) steric hindrance of the methyl groups for the adsorption of the molecule hydrogenated, (b) steric hindrance of the methyl groups for the cleavage of the C—S bond, (c) the fact that only one atom of H is available for the cleavage of the C—S bond, and (d) an effect of the methyl group on the acidity of the H atom involved in the elimination process.

The mechanism of elimination E_2 (β-elimination) is described as following: a group $S:^-$ (nucleophile) subtracts a proton from the molecule with sulfur atom and the leaving group is the S atom from that molecule (**Figure 3**).

The favorable configuration for an elimination E_2 is to have both the sulfur atom and the hydrogen-β atom interacting with the surface of the active phase ($Mo(W)S_2$) at the same moment (**Figure 4a**). Then, the methyl group can hinder the process of elimination by blocking either the sulfur atom or the hydrogen-β atom as it approaches to catalytic center (**Figure 4b**). Moreover, the methyl group in 4,6-DMDBT molecule can also cause the hydrogen-β atom involved in the elimination process to be less acidic than the DBT molecule (**Figure 4c**). On the other hand, in the case of 4,6-DMDBT molecule, only one hydrogen atom is available for the elimination instead of two H atoms as occurs in DBT molecule. All of these factors can lower the reactivity of the 4,6-DMDBT compared to the DBT molecule.

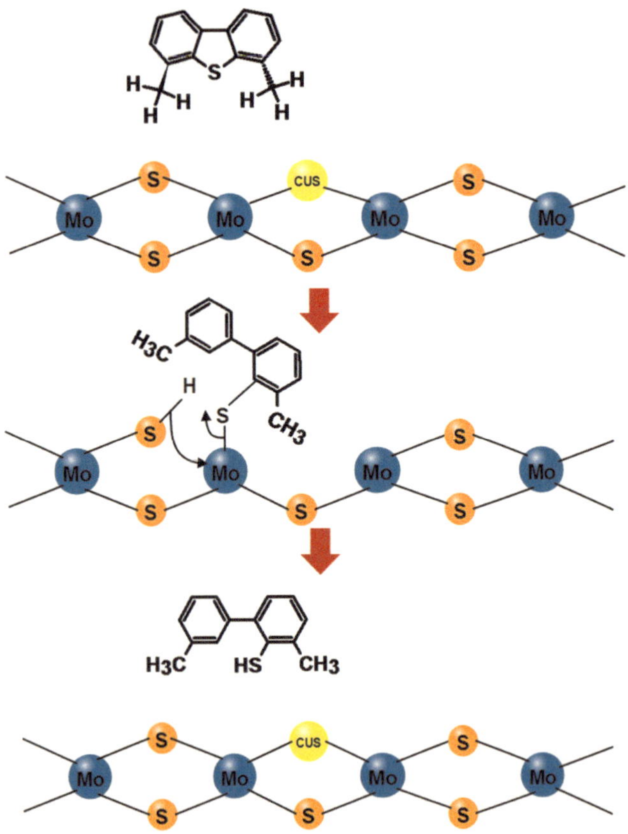

Figure 2.
Mechanism of the cleaving of the C—S bond in the DDS route.

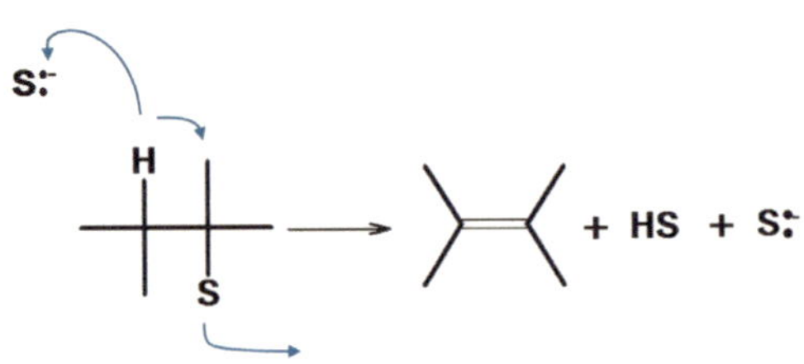

Figure 3.
Mechanism of elimination E2 (β-elimination).

2.2 Active phase models

In the literature, there are some models that try to explain the structure and operation of the active phase. The models proposed by Daage and Chianelli [20] and Chianelli et al. [21] have named the sites located in the upper and lower edges of the crystal as "Rim site," which are reactive to the reactions of HYD and the break of the C—S bond. While the "edge" sites are active only in the cleaving of the C—S. **Figure 5** depicts the location of these sites in the MoS_2 crystal.

Ramos et al. [22] have also showed for unsupported systems the existence of strong electron donation from Co to Mo and an enhanced metallic character associated to the Co_9S_8/MoS_2 interface. Berhault et al. [23] studied the structural role of cobalt, and the influence of support interactions on the morphology and catalytic properties of Mo and CoMo catalysts supported on alumina and silica.

On the other hand, another model called of the mixed phase "Co(Ni)-Mo(W)-S" combines studies of tunneling microscopy (STM) with calculations of density functional theory (DFT), identifying an area with high electron density in the upper part of the MoS_2 crystal that was called "BRIM site," which have metal properties capable of efficiently carrying out the hydrogenation reactions [19, 24–26]. In the present chapter, the mixed phase model was used, since it is the most widely

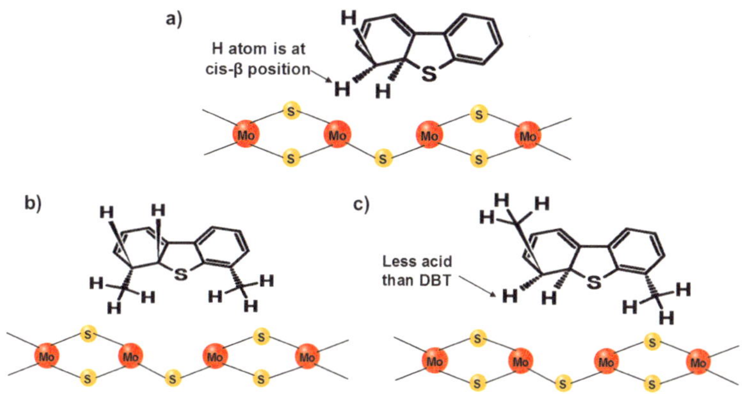

Figure 4.
Approaching of: (a) DBT, (b) methyl group of the 4,6-DMDBT, and (c) hydrogen atom of the 4,6-DMDBT molecule to the catalytic center.

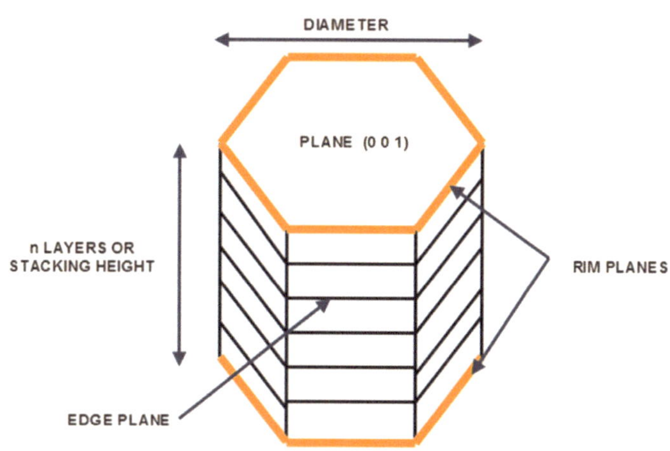

Figure 5.
Model "Rim-edge" [21].

utilized in the literature. Based on this model, the catalytic properties of the sulfur vacancies at the edges of $Mo(W)S_2$ crystallites are strongly enhanced by the close presence of a promoter atom (Co or Ni) in the so-called Co(Ni)-Mo(S)-S structures [23, 26–28]. It has been reported that there are two types of structures that involve molybdenum or tungsten Co(Ni)-Mo(W)-S crystal, these structures were named types I and II [23, 27–29]. Type I structure has a strong interaction in the γ-alumina, since there is the presence of Mo-O-Al linkages and presents poor sulfidation. Type II structures are characterized by a weaker interaction with the γ-alumina, allowing to be full sulfided and exhibits high HDS activity. Therefore, it is important to find a way to weaken the interaction between the active phase and the support. One option is to use supports that present weak interaction with the active phase such as silica or carbon. On these supports (silica or carbon) "multilayer" MoS_2 structures are generated, thus the superior crystallites in the structure have a lower interaction with the support and form type II structures, which means that stacked or multilayer structures are type II and that single layer structures are type I, as shown in the **Figure 6**.

Nevertheless, it is also true that single layer structures that are type II can be obtained through a complete sulfidation of the oxidized phase and weaken the electronic interaction with the support. **Figure 7** illustrates the aforementioned.

The choice of a suitable support material is often dictated by the process conditions in which a catalyst has to operate. Although, one of the key features of a

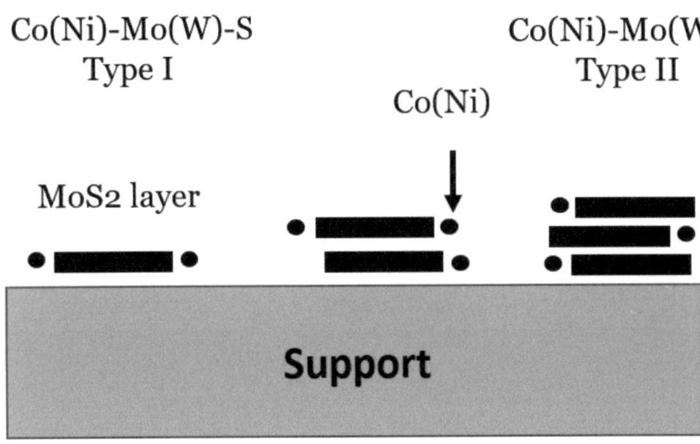

Figure 6.
Structures Co(Ni)-Mo(W)-S types I and II.

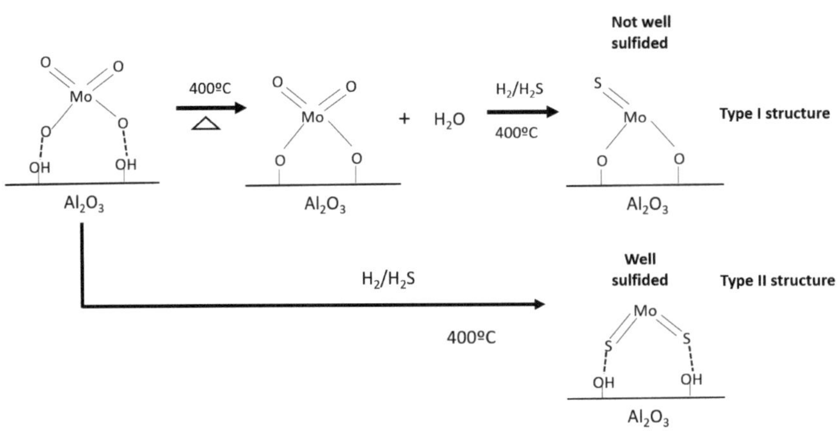

Figure 7.
Sulfidation reaction in Mo/Al2O3 catalysts.

support for HDS catalysts is the presence of a large specific area in which the active phase $Mo(W)S_2$ presents a very high dispersion and/or present weak interaction support—active phase to generate more active structures (type II). Therefore, it is important to find a way to weaken the interaction between the active phase and the support. Therefore, the purpose of the next section is to make an overview of the recent investigations into the role of silicon in the generation of the type II structures (weak metal-support interaction), which present better HDS conversion.

3. HDS catalysts supported on mesoporous silicates

The mesoporous silicates such as MCM-41 and SBA-15 have received great attention in the last decades due to their excellent properties as catalyst support in hydrotreating reactions [30–36]. These catalysts have reported a high catalytic activity in the HDS of DBT than their counterpart supported on alumina. In the case of MCM-41, its low structural stability has limited its industrial use as support. It is also reported that the addition of Al_2O_3 stabilizes the structure of the MCM-41, but do not achieve the promoting effect of Ni [37].

In order to obtain a deep insight into how the preparation conditions of the MCM-41 have influence on the performance of the catalyst, Hernandez Cedeño et al. [38] did a detail study of UV-vis spectroscopy in order to find out how the pH (7 and 9) during the preparation of MCM-41 (MCM41) and Al_2O_3 (Al) supports as well as the effect of varying the molar ratio of Si/Al (10, 25, and 50) affect the coordination of the supported metals, and thus the metal-support interaction. The catalysts NiW were evaluated in the HDS of DBT. For the catalysts, W/Al_2O_3 at different pH presented well-defined bands at 423, 720–722 nm, which can be related to nickel in octahedral coordination [39, 40].

On the other hand, the catalysts NiW/MCM41 prepared at pH 7 with different Si/Al (10, 25 and 50) molar ratio presented bands in the range of 710–730 nm ascribed to octahedral nickel species [11, 49–51]. But also a band between 807 and 818 nm were detected, which was associated to nickel species with octahedral symmetry distorted; this latter may be related to an interaction Ni-W [41, 42]. In addition, the catalysts prepared at pH 9 bands at 804 and 820 nm ascribed to nickel with octahedral distortion were also detected similarly to the catalyst synthesized at pH 7 [43].

At pH of 9, the NiW/AMS50 prevails the band at 818 nm, related to nickel with distorted octahedral symmetry, while for the NiW/AMS10 using the same pH, no octahedral species were observed. An opposite behavior was observed with the pH 7 catalyst, which means the catalyst NiW/AMS10 presents more define bands related to nickel species in octahedral coordination with respect to the NiW/AMS 50.

Then, the catalyst with the highest reaction rate at pH 7 and pH 9 was the NiW/AMS10 and NiW/AMS50 catalysts, respectively. Therefore, there is a trend between the HDS activity and the bands in the range of 804–820 nm as well as the bands between 710 and 730 nm, which are related to the interaction of Ni-W and Ni species in octahedral coordination, respectively. Meanwhile, on the catalyst NiW/Al_2O_3 prevailed the well-defined band between 720 and 775 nm associated to Ni2+ in octahedral symmetry, which is caused by the interaction between the metal and promoter, and leads to the formation of species NiWS type II, resulting in a better conversion of the DBT.

Although the incorporation of Al in the MCM41 framework improved the hydrothermal stability, the reaction rate decreases. According to the results of that work, the incorporation of Al in the MCM41 framework has a negative effect on the catalyst performance. On the other hand, in the NiW/AMS catalyst, both the pH and the amount of alumina affect the reaction rate. In the case of the catalysts impregnated at pH 7, the highest reaction rate was achieved with the catalyst that contains the highest amount of alumina in its structure (NiW/AMS10). On the contrary for the catalysts impregnated at pH 9, the NiW/AMS50 catalyst (lowest amount of alumina) presented the highest reaction rate of this series.

Continuing with the influence of the preparation conditions of the support with Si on the performance of the HDS catalysts in terms of metal-support interaction, Gómez-Orozco et al. [44] analyzed the effect of modify the SBA-15 support with Ti on its physicochemical properties and its sulfidation behavior using NiMoW/SBA-15 catalysts in the HDS reaction of DBT. The amounts of Ti4+ ions incorporated during the direct synthesis of the SBA-15 support were varied using an Si/Ti molar ratio of 60, 40, and 20, and the nominal metals loading were 3.84, 13.83, and 17.33 wt% of Ni, Mo, and W, respectively. The supported catalysts were labeled as CAT/S15, CAT/60TiS15, CAT/40TiS15, and CAT/20TiS15 in agreement with the nominal Si/Ti ratio of 60, 40, or 20, respectively.

The coordination of Ni, Mo, and W ions was analyzed by UV-vis-DRS. It was observed that all the samples present a strong band in the range of 210–280 nm, which is ascribed to Mo and W ions in tetrahedral coordination, such as Mo(W) $O4^{2-}$. The intensity of this band (210–280 nm) was observed to follow the next trend: CAT/40TiS15 > CAT/20TiS15 > CAT/S15 > CAT/60TiS15. The catalysts CAT/40TiS15 and CAT/60TiS also showed an intense band at about 350 nm, which is related to Mo or W ions with octahedral coordination [45]. Furthermore, these catalysts present a band at 750 nm assigned to Ni^{2+} ions in octahedral coordination. In general, the population of octahedral W species and the W species in tetrahedral coordination was increased significantly upon Ti incorporation into S15. In addition, the incorporation of Ti did not decrease the catalytic activity in the HDS of DBT reaction except for Si/Ti = 40. The UV-vis analysis of the catalyst CAT/40TiS15 indicates a higher amount of tetrahedral species than its counterpart CAT/60TiS15. The lower amount of Ti in the catalyst (CAT/60TiS15) presented the highest hydrogenation capability among the catalysts studied (HYD/DDS = 0.81) (**Table 1**) despite the main route of DBT reaction was the direct desulfurization (DDS) pathway. Hence, the superior activity for CAT/60TiS15 and CAT/S15 samples was related to a higher dispersion of Mo(W)S2 phase and a lower amount of tetrahedral species, which are not easy to reduce and sulfide (type I structure), which means that the catalysts CAT/60TiS15 and CAT/S15 possess a higher population of Mo(W)S2 phase with type II structure.

The incorporation of Ti into the structure of the SBA-15 affected the catalytic properties of the sulfide catalysts. However, the Ti effect depends on its loading. In this study, the moderate Ti loading (Si/Ti = 40 molar ratio) was observed and had a negative effect in the catalytic activity, presenting the lower DBT conversion

Catalysts	DDS	HYD				DBT
	BF (%)	CBH (%)	BCH (%)	THDBT (%)	HYD/DDS	Conversion (%)
CAT/S-15	22.34	13.47	1.11	3.08	0.79	91.57
CAT/60Ti-S15	22.16	13.05	1.05	3.75	0.81	96.86
CAT/40Ti-S15	23.99	12.3	0.1	3.6	0.65	80.15
CAT/20Ti-S15	22.86	13.23	0	3.58	0.75	82.36

THDBT, tetrahydrodibenzothiophene; BF, biphenyl; CBH, cyclohexylbenzene; and BCH, bicyclohexyl; calculated for batch reactor operating at T = 320°C and PH_2 = 800 psi for 5 h.

Table 1.
DBT conversion and selectivity HYD/DDS (at 40% of DBT conversion) [39].

(**Table 1**). One reason for the poor activity in the catalysts CAT/40TiS15 is the higher amount of tetrahedral species compared to their counterparts. In contrast, the catalyst CAT/60TiS15 (Si/Ti = 60 molar ratio) presented a lower amount of the tetrahedral species. Tetrahedral species are difficult to be reduced and sulfide, and therefore are not susceptible to develop the HDS active sites. Then, CAT/60TiS15 exhibited the highest HDS activity (96.98% of DBT conversion), the better performance for this catalyst is due to the small amount of tetrahedral species, the low staking, and high dispersion of Mo(W)S2 phases. As it is observed, there is an optimal relationship of Si/Ti molar ratio in order to improve the HDS performance. Furthermore, the incorporation of certain amount of Ti into the structure of SBA-15 generates a high dispersion of the active phase, and a large number of structure type II.

4. HDS catalysts supported on SiO2-Al2O3

As mentioned above, the type I structure with poor sulfidation has strong Mo-O-Al linkage with γ-alumina and presents low activity, whereas the type II structures with full sulfidation possess weak interaction with γ-alumina and exhibit high HDS activity. The surface modification of alumina with silica is an efficient way to weaken the metal-support interaction. Sanchez-Minero et al. [46] studied the effect of incorporate SiO2 with a nominal loading of 10 wt% (SAC 10) onto the surface of alumina in NiMo/Al2O3-SiO2(x) catalysts for the hydrotreatment of mixtures of 4,6-DMDBT-naphthalene-carbazole. An infrared analysis of the hydroxyl region was carried out, characteristics bands of Al2O3 hydroxyl groups were observed at 3790, 3775, 3740, 3730 and 3680 cm^{-1} [47], and it can be noted from the IR spectrum of the alumina support (SAC). The most basic hydroxyl groups in alumina give rise to an IR band at 3775 cm^{-1}. When silica is incorporated to alumina (SAC 10), some changes in the bands intensity are observed. A new band localized in the region at 3725–3750 cm^{-1} appears, which is assigned to isolated silanol groups [48, 49]. Furthermore, the bands corresponding to the most basic hydroxyl groups (3775 cm^{-1}) disappear. This behavior indicates that modifying the alumina surface with SiO2 eliminates the most basic hydroxyl groups in alumina promoting that the sulfided NiMoSAC10 catalyst presented highly stacked MoS2 crystallites with more than two layers (type II structure). These sites favor the hydrogenation route, being more active for 4,6-DMDBT HDS.

Recently, Romero-Galarza et al. [50] carried out a systematic study of the change in activity, selectivity, dispersion, sulfidation, and extent of promotion for CoMo and NiMo HDS catalysts supported on Al2O3 and SiO2/Al2O3. They found for CoMo and NiMo catalysts that the grafting of the surface of alumina support with a 4.0 wt% of silica was enough to eliminate the most basic hydroxyl groups bonded to tetrahedral aluminum (IR band at 3767 cm^{-1}), and thus, this induces to have a higher proportion of Mo and Co(Ni) in octahedral coordination (DRS-UV-vis results), resulting in a better catalytic performance in the HDS of 4,6-DMDBT.

It was also found that the extent of promotion, determinated by the XPS ratio of NiMoS/NiT, is larger for the Ni-promoted catalysts than for Co-promoted catalyst, which is in line with the fact that NiMo catalysts can incorporate the Ni promoter on three different edges of a dodecagonal NiMoS particle, in contrast to CoMo catalysts where the copromoter is incorporated only on the sulfur edge of a hexagonal CoMoS cluster [26, 27]. The origin of the better performance of the NiMoSAC catalyst over their alumina-supported counterparts, NiMoAl and CoMoAl, seems to be mainly related to the higher extent of promotion and sulfidation achieved in the catalysts with SiO2 (type II structures). This fact is reflected with a good

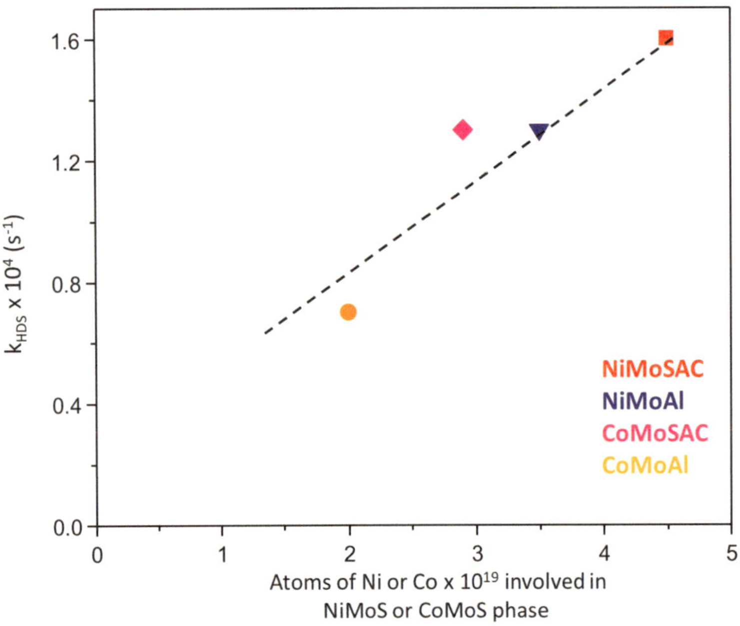

Figure 8.
Relationship between global HDS rate constant vs. Ni or Co atoms involved in the Co(Ni)MoS phase.

correlation between the degree of promotion with the hydrodesulfurization rate constant displayed in **Figure 8**, and agrees with the literature reports that indicate that promotion favors the appearance or brighter (more metallic) brim site, which can perform hydrogenating reactions [51], which is the main route for the HDS of 4,6-DMDBT.

On the other hand, Xu et al. [11] prepared a novel NiMo/SiO2-Al2O3 catalysts with the improved stacking and good dispersion of supported active phase via gemini surfactant-assisted synthesis. In this method, polymolybdates anions were transformed into gemini surfactant-linked Mo precursor (GSMP), dispersing Mo species well and weakening the strong Mo-support interaction. The GSMP-based NiMo/SiO2-Al2O3 (NiMo-GSHD) catalyst presents higher activity for the HDS of 4,6-DMDBT than its counterparts prepared via impregnation (NiMo-IM) and the cetyltrimethylammonium bromide-assisted hydrothermal method (NiMo-CTHD). To understand their different activities (**Table 2**), the HDS activities of the catalysts were correlated with the structure of their metal phase.

The reason for the higher HDS activity of 4,6-DMDBT (**Table 2**) exhibited in the catalyst NiMo-GSHD is related to the greater MoS2 dispersion, a superior average stacking number determinated by HRTEM and the higher extent of promotion (NiMoS) calculated by XPS and by NO-IR characterization, thus generating more Ni-Mo-S active sites with sufficient brim sites (type II structures). Such that, the prehydrogenation activity of NiMi-GSHD for 4,6-DMDBT with steric hindrance is markedly improved. The prehydrogenated products (4,6-THDMDBT and 4,6-HHDMDBT) without steric hindrance are much easier to be desulfurization via hydrogenolysis on the edge sites of Ni-Mo-S phases than initial 4,6-DMDBT. Therefore, NiMo-GSHD, with more edge sites due to its better metal dispersion, possesses higher 3,3´-MCHT selectivity than NiMo-IM and NiMo-CTHD.

According to the results showed in this section, the use of supports SiO2-Al2O3 results in an improvement in the performance of the catalysts in the HDS reaction of the 4,6-DMDBT molecule. Grafting SiO_2 on the surface of γ-alumina generates two main effects. Increasing the extent of sulfidation and promotion, generating

Catalysts	k_{HDS}[a]	TOF[b] × 10^4	Product ratio[c]	
	(10^{-7} $molg^{-1}s^{-1}$)	(s^{-1})	(TH + HH)/MCHT	MCHT/DMDBP
Mo-IM	0.32	1.72	0.33	1.85
Mo-CTHD	0.47	2.24	0.3	2.41
Mo-GSHD	0.7	2.83	0.27	3.07
NiMo-IM	3.16	5.62	0.16	2.04
NiMo-CTHD	4.18	7.34	0.13	2.72
NiMo-GSHD	5.78	9.21	0.11	3.67

TH, tetrahydrodimethylbenzothiophene; HH, hexahydrodimethyldibenzothiophene; MCHT, dimethylbicyclohexyl; and DMDBP, dimethylbiphenyl.
[a]*Calculated with the 4,6-DMDBT conversion at about 30%.*
[b]*Number of the reacted 4,6-DMDBT molecules per second and per Mo atom at the edge surface.*
[c]*Determinated at about 50% of the total 4,6-DMDBT conversion by changing liquid hourly space velocity.*

Table 2.
HDS results of 4,6-DMDBT on different catalysts [47].

structures Co(Ni)-Mo-S type II, unlike the catalysts supported in γ-alumina where the structures type I predominate, which are not well sulfided due to a strong metal-support interaction. On the other hand, a different Mo precursor is used (gemini surfactant-linked) in conjunction with the use of a mixed Al2O3-SiO2 support with a composition of 96.4 and 3.6 wt%, respectively, resulting in the formation of sulfide molybdenum (MoS2) crystals with higher stacking, generating the so-called Ni-Mo-S type II, which are more active in HDS of 4,6-DMDBT molecule.

5. Conclusions

The metal-support interaction is one of the most important parameters in the design of HDS catalysts. The use of silicon in the preparation of HDS catalyst support has been showed to weak the metal interaction, generating type II Co(Ni)-Mo(W)-S structures, which are characterized by: (i) a complete sulfidation of the oxidized phase, weakening the electronic interaction with the support or (ii) stacked structures so that the upper crystallites in the structure Mo(W)S2 have a low interaction with the support. The so-called type II Co(Ni)-Mo(W)-S structures are more active than the partially sulfide type I. The use of mesoporous silicates such as MCM-41 and SBA-15 has been proposed in the literature, with the intention to increase the active phase, acidity, the type II structures, and the dispersion of the active phase (Mo(W)S2). In the case of the MCM-41 support, the preparation conditions such as pH and the Si/Al molar ratio increase the number of oxidized species in octahedral coordination, which are precursors of the type II structures. However, these materials did not show better HDS performance in the DBT molecule compared with the alumina support. It would be very useful to evaluate this type of catalyst support (MCM-41) in a molecule more refractory to HDS, such as 4,6-DMDBT, and see if it is possible to increase the catalytic performance compared to the catalyst supported in alumina, since the 4,6-DMDBT molecule is more sensitive to the geometry of the Co(Ni)-Mo(W)-S structure than the DBT molecule. On the other hand, SBA-15 mesoporous silicate doped with a certain amount of Ti (Si/Ti = 60 molar ratio) showed to improve the catalytic performance in the HDS of the DBT molecule, through generating a greater population of type II structures; therefore, these materials are useful as HDS catalyst support.

Grafting SiO2 on the surface of γ-alumina or the use of mixed support (Al2O3-SiO2) are other alternatives to promote the formation of so-called type II Co (Ni)-Mo(W)-S structures. It has been reported that a small amount of SiO2 is enough to weaken the metal-support interaction and thus increase the extent of sulfidation and level promotion, which are fundamental parameters to improve the performance of HDS catalysts. So, it can be concluded that the use of silicon in the preparation of HDS catalyst support is a promising alternative to get better performance in HDS catalysts.

Author details

Denis A. Cabrera-Munguia[1], Lucero Rosales-Marines[2], Anilu Rubio-Ríos[2], Lorena Farías-Cepeda[2] and Adolfo Romero-Galarza[2*]

1 Chemical Engineering Faculty, Michoacana de San Nicolás de Hidalgo University, Morelia, Michoacán, Mexico

2 Chemical Engineering Department, Faculty of Chemical Sciences, Autonomous University of Coahuila, Saltillo, Coahuila, Mexico

*Address all correspondence to: a_romero@uadec.edu.mx

References

[1] Rangarajan S, Mavrikakis M. On the preferred active sites of promoted MoS2 for hydrodesulfurization with minimal organonitrogen inhibition. ACS Catalysis. 2016;7:501-509. DOI: 10.1021/acscatal.6b02735

[2] Fu W, Zhang L, Tang T, Ke Q, Wang S, Hu J, et al. Extraordinarily high activity in the hydrodesulfurization of 4,6-dimethyldibenzothiophene over Pd supported on mesoporous zeolite Y. Journal of the American Chemical Society. 2011;**133**:15346-15349. DOI: 10.1021/ja2072719

[3] Egorova M, Prins R. Competitive hydrodesulfurization of 4,6-dimethyldibenzothiophene, hydrodenitrogenation of 2-methylpyridine, and hydrogenation of naphthalene over sulfide NiMo/γ-Al2O3. Journal of Catalysis. 2004;**224**:278-287. DOI: 10.1016/j.jcat.2004.03.005

[4] Recursos energéticos globales, Encuesta 2013: Resumen. World Enegy council. Available from: https://www.worldenergy.org/wp-content/uploads/2014/04/Traduccion-Estudio-Recursos-Energeticos1.pdf

[5] Stanislaus A, Marafi A, Rana SM. Recent advances in the science and technology of ultra low sulfur diesel (ULSD) production. Catalysis Today. 2010;**153**:1-68. DOI: 10.1016/j.cattod.2010.05.011

[6] ENI. 2016. O&G World oil and gas review 2016. Avaliable from: https://www.eni.com/docs/en_IT/enicom/company/fuel-cafe/WOGR-2016.pdf

[7] Van Haandel L, Bremmer M, Kooyman PJ, Van Veen JAR, Weber T, Hensen EJM. Structure-activity correlations in hydrodesulfurization reactions over Ni-promoted MoxW(1-x)S2/Al2O3 catalysts. ACS Catalysis. 2015;**5**:7276-7287. DOI: 10.1021/acscatal.5b01806

[8] Vít Z, Gulková D, Kaluza L, Kupcik J. Pd-Pt catalysts on mesoporous SiO2-Al2O3 with superior activity for HDS of 4,6-dimethyldibenzothiophene: Effect of metal loading and support composition. Applied Catalysis B: Environmental. 2015;**179**:44-53. DOI: 10.1016/j.apcatb.2015.04.057

[9] Xiangchen F, Rong G, Chengmin Y. The development and application of catalysts for ultra-deep hydrodesulfurization of diesel. Chinese Journal of Catalysis. 2013;**34**:130-139. DOI: 10.1016/S1872-2067(11)60506-8

[10] Sh R, Li J, Feng B, Wang Y, Zhang W, Wen G, et al. Catalysis Today. 2015;**263**:136-140. DOI: 10.1016/j.cattod.2015.06.023

[11] Xu J, Huang T, Fan Y. Highly efficient NiMo/SiO2-Al2O3 hydrodesulfurization catalyst prepared from gemini surfactant-dispersed Mo precursor. Applied Catalysis B: Environmental. 2017;**203**:839-850. DOI: 10.1016/j.apcatb.2016.10.078

[12] López-Benítez A, Berhault G, Guevara-Lara A. Addition of manganese to alumina and its influence on the formation of supported NiMo catalysts for dibenzothiophene hydrodesulfurization application. Journal of Catalysis. 2016;**344**:59-76. DOI: 10.1016/j.jcat.2016.08.015

[13] Gates BC, Topsøe H. Reactives in deep catalytic hydrodesulfurization: Challenges, opportunities, and the importance of 4-methyldibenzothiophene and 4,6-dimethyldibenzothiophene. Polyhedron. 1997;**16**:3213-3217. DOI: S0277-5387(97)00074-0

[14] Song C. An overview of new approaches to deep desulfurization for

ultra-clean gasoline, diesel and jet fuel. Catalysis Today. 2003;**86**:211-263. DOI: 10.1016/S0920-5861(03)00412-7

[15] Babich IV, Moulijn JA. Science and technology of novel processes for deep desulfurization of oil refinery streams: A review. Fuel. 2003;**82**:607-631. DOI: S0016–2361(0 2)00324–1

[16] Li X, Wang A, Egorova M, Prins R. Kinetics of the HDS of 4,6-dimethyldibenzothiophene and its hydrogenated intermediates over sulfide Mo and NiMo on γ-Al2O3. Journal of Catalysis. 2007;**250**:283-293. DOI: 10.1016/j.jcat.2007.06.005

[17] Bataille F, Louis Lemberton J, Michaud P, Perot G, Vrinat M, Lemaire M, et al. Alkyldibenzothiophenes hydrodesulfurization-promoter effect, reactivity, and reaction mechanism. Journal of Catalysis. 2000;**191**:409-422. DOI: 10.1006/jcat.1999.2790

[18] Moses PG, Hinneemann B, Topsøe H, Nørskov JK. The hydrogenation and direct desulfurization reaction pathway in thiophene hydrodesulfurization over MoS2 catalysts at realistic conditions: A density functional study. Journal of Catalysis. 2007;**248**:188-203. DOI: 10.1016/j.jcat.2007.02.028

[19] Lauritsen JV, Nyberg M, Nørskov JK, Clausen BS, Topsøe H, Laegsgaard E, et al. Hydrodesulfurization reaction pathways on MoS2 nanoclusters revealed by scanning tunneling microscopy. Journal of Catalysis. 2004; **224**:94-106. DOI: 10.1016/j. jcat.2004.02.009

[20] Daage M, Chianelli RR. Structure-Function in molybdenum sulfide catalysts: The "rim-edge" model. Journal of Catalysis. 1994;**149**:414-427. DOI: 10.1006/jcat.1994.1308

[21] Chianelli RR, Siadati MH, Perez De la Rosa M, Berhault G, Wilcoxon JP, Bearden R Jr, et al. Catalytic properties of single layers of transition metal sulfide catalytic materials. Catalysis Reviews. 2006;**48**:1-41. DOI: 10.1080/01614940500439776

[22] Ramos M, Berhault G, Ferrer DA, Torres B, Chianelli RR. Catalysis Science δ Technology. 2012;**2**:164-178. DOI: 10.1039/c1cy00126d

[23] Berhault G, Perez De la Rosa M, Mehta A, Yácaman MJ, Chianelli RR. The single-layered morphology of supported MoS2-based catalysts—The role of the cobalt promoter and its effects in the hydrodesulfurization of dibenzothiophene. Applied Catalysis A: General. 2008;**345**:80-88. DOI: 10.1016/j.apcata.2008.04.034

[24] Walton AS, Lauritsen JV, Topsoe H, Besenbacher F. MoS2 nanoparticles morphologies in hydrodesulfurization catalysis studied by scanning tunneling microscopy. Journal of Catalysis. 2013; **308**:306-318. DOI: 10.1016/j.jcat.2013.08.017

[25] Kibsgaard J, Tuxen A, Knudsen KG, Brorson M, Topsøe H, Laegsgaard E, et al. Comparative atomic-scale analysis of promotional effects by late 3-d-transition metals in MoS2 hydrotreating catalysts. Journal of Catalysis. 2010;**272**: 195-203. DOI: 10.1016/j.jcat.2010.03.018

[26] Lauritsen JV, Kibsgaard J, Olsen GH, Moses PG, Hinnemann B, Helveg S, et al. Location and coordination of promoter atoms in Co and Ni promoted MoS2-based hydrotreating catalysts. Journal of Catalysis. 2007;**249**:220-233. DOI: 10.1016/j.jcat.2007.04.013

[27] Topsøe H. The role of Co-Mo-S type structures in hydrotreating catalysts. Applied Catalysis A: General. 2007;**322**: 3-8. DOI: 10.1016/j.apcata.2007.01.002

[28] Brorson M, Carlsson A, Topsøe H. The morphology of MoS2, WS2, Co-Mo-S, Ni-Mo-S and Ni-W-S nanoclusters in

hydrodesulfurization catalysts revealed. Catalysis Today. 2007;**123**:31-36. DOI: 10.1016/j.cattod.2007.01.073

[29] Hinnemann B, Nørskov JK, Topsøe H. A density functional study of the chemical differences between type I and type II MoS2-based structures in hydrotrating catalysts. The Journal of Physical Chemistry B. 2005;**109**:2245-2253. DOI: 10.1021/jp048842y

[30] Vutolkina AV, Glotov AP, Zanina AV, Mahmutov DF, Maksimov AL, Karakhanov EA. Mesoporous Al-HMS and Al-MCM-41 supported Ni-Mo sulfide catalysts for HYD and HDS via in situ hydrogen generation through a WGSR. Catalysis Today. 2019;**329**:156-166. DOI: 10.1016/j.cattod.2018.11.030

[31] Zhou W, Wei Q, Zhou Y, Liu M, Ding S, Yang Q. Hydrodesulfurization of 4,6-dimethyldibenzothiophene over NiMo sulfide catalysts supported on meso-microporous Y zeolite with different mesopore sizes. Applied Catalysis B: Environmental. 2018;**238**: 212-224. DOI: 10.1016/j.apcatb.2018.07.042

[32] Suresh C, Pérez Cabrera L, Aliaga JA, Díaz de León JN, Zepeda TA, Fuentes S, et al. Formation of Co-promoted MoS2 fullerene-like nanostructures on SBA-15 as effective hydrodesulfurization catalysts. Catalysis Letters. 2017;**147**:46-57. DOI: 10.1007/s10562-016-1936-x

[33] Alonso-Núñez G, Bocarando J, Huirache-Acuña R, Álvarez-Contreras L, Huang ZD, Bensch W, et al. Influence of the activation atmosphere on the hydrodesulfurization of Co-Mo/SBA-15 catalysts prepared from sulfur-containing precursors. Applied Catalysis A: General. 2012;**419-420**:95-101. DOI: 10.1016/j.apcata.2012.01.015

[34] Zepeda TA, Infantes-Molina A, Díaz de Léon JN, Fuentes S, Alonso-Núñez G, Torres-Otañez G, et al. Hydrodesulfurization enhancement of heavy and light S-hydrocarbons on NiMo/HMS catalysts modified with Al and P. Applied Catalysis A: General. 2014;**484**:108-121. DOI: 10.1016/j.apcata.2014.06.033

[35] Ren J, Wang A, Li X, Chen Y, Liu H, Hu Y. Hydrodesulfurization of dibenzothiophene catalyzed by Ni-Mo sulfides supported on a mixture of MCM-41 and HY zeolite. Applied Catalysis A: General. 2008;**344**:175-182. DOI: 10.1016/j.apcata.2008.04.017

[36] Zepeda TA, Fierro JLG, Pawlec B, Nava R, Klimova T, Fuentes GA, et al. Synthesis and characterization of Ti-HMS and CoMo/Ti-HMS oxide materials with varying Ti content. ACS Chemistry of Materials. 2005;**17**:4062-4073. DOI: 10.1021/cm0500051

[37] Silva-Rodrigo R, Hernández-Lopez F, Martinez-Juarez K, Castillo-Mares A, Melo Banda JA, Olivas-Sarabia A, et al. Synthesis, characterization and catalytic properties of NiMo/Al2O3-MCM-41 catalyst for dibenzothiophene hydrodesulfurization. Catalysis Today. 2008;**130**:309-319. DOI: 10.1016/j.cattod.2007.10.001

[38] Hernandez Cedeño G, Silva-Rodrigo R, Guevara-Lara A, Melo-Banda JA, Reyes de la Torre AI, Morteo Flores F, et al. Role of the Si/Al molar ratio and pH in NiW/MCM41-Al2O3 catalysts for HDS of DBT. Catalysis Today. 2016;**271**: 64-79. DOI: 10.1016/j.cattod.2015.10.024

[39] Priecel P, Kubicka D, Capek L, Bastl Z, Rysanek P. The role of Ni species in the deoxygenation of rapeseed oil over NiMo-alumina catalysts. Applied Catalysis A: Gerneral. 2011;**397**:127-137. DOI: 10.1016/j.apcata.2011.02.022

[40] López-Fonseca R, Jiménez-González C, De Rivas B, Gutiérrez-Ortiz J. Partial oxidation of methane to syngas on bulk $NiAl_2O_4$ catalyst. Comparison

with alumina supported nickel, platinum and rhodium catalysts. Applied Catalysis A: General. 2012;**437-438**:53-62. DOI: 10.1016/j.apcata.2012.06.014

[41] Cruz-Perez AE, Guevara-Lara A, Morales-Ceron JP, Alvarez-Hernandez A, De los Reyes JA, Massin L, et al. Ni and W interactions in the oxide and sulfide states on an Al2O3-TiO2 support and their effects on dibenzothiophene hydrodesulfurization. Catalysis Today. 2011;**172**:203-208. DOI: 10.1016/j.cattod.2011.03.027

[42] Díaz de Léon JN, Picquart M, Massin L, Vrinat M, De los Reyes JA. Hydrodesulfurization of sulfur refractory compounds: Effects of gallium as an additive in NiWS/γ-Al2O3 catalysts. Journal of Molecular Catalysis A: Chemical. 2012;**363-364**:311-321. DOI: 10.1016/j.molcata.2012.07.006

[43] Cervantes-Gaxiola ME, Arroyo-Albiter M, Maya-Yesca R, Rico-Cerda JL, Guevara-Lara A, Espino-Valencia J. Synthesis, characterization and catalytic activity during hydrodesulphurization of dibenzothiophene of NiMoW catalysts supported on Al-Ti mixed oxides modified with MgO. Fuel. 2012;**100**:57-65. DOI: 10.1016/j.fuel.2011.12.040

[44] Gómez-Orozco SY, Huirache-Acuña R, Pawlec B, Fierro JLG, Rivera-Muñoz EM, Lara-Romero J, et al. Characterization and HDS performance of sulfided NiMoW catalysts supported on mesoporous titania-modified SBA-15. Catalysis Today. 2018;**305**:152-161. DOI: 10.1016/j.cattod.2017.08.009

[45] Lee DK, Lee HT, Lee IC, Park SK, Bae SY, Kim CH, et al. W-incorporated CoMo/γ-Al2O3 hydrodesulfurization catalyst II. Characterization. Journal of Catalysis. 1996;**159**:219-229. DOI: 0021-9517/96

[46] Sánchez-Minero F, Ramírez J, Gutiérrez-Alejandre A, Fernández-Vargas C, Torres-Mancera P, Cuevas-Garcia R. Analysis of the HDS of 4,6-DMDBT in the presence of naphthalene and carbazole over NiMo/Al2O3-SiO2 (x) catalysts. Catalysis Today. 2008;**133-135**:267-276. DOI: 10.1016/j.cattod.2007.12.018

[47] Digne M, Sautet P, Raybaud P, Euzen P, Toulhoat H. Hydroxyl groups on γ-alumina surfaces: A DFT study. Journal of Catalysys. 2002;**211**:1-5. DOI: 10.1006/jcat.2002.3741

[48] Bianchini D, Zimnoch dos Santos JH, Uozumi T, Sano T. Characterization of MAO-modified silicas. Journal of Molecular Catalysis A: Chemical. 2002;**185**:223-235. DOI: S1381-1169(02)00047-X

[49] Shah P, Ramaswamy AV, Lazar K, Ramaswamy V. Synthesis and characterization of tin oxide-modified mesoporous SBA-15 molecular sieves and catalytic activity in trans-esterification reaction. Applied Catalysis A: General. 2004;**273**:239-248. DOI: 10.1016/j.apcata.2004.06.039

[50] Romero-Galarza A, Ramírez J, Gutiérrez-Alejandre A, Solís-Casados DA. Relevant changes in the properties of Co(Ni)Mo/Al2O3 HDS catalysts modified by small amounts of SiO2. Journal of Material Research. 2018;**33**:3570-3579. DOI: 10.1557/jmr.2018.263

[51] Basenbacher F, Brorson M, Clausen BS, Helveg S, Hinnemann B, Kibsgaard J, et al. Recent STM, DFT and HAADF-STEM studies of sulfide-based hydrotreating catalysts: Insight into mechanistic, structural and particle size effects. Catalysis Today. 2008;**130**:86-96. DOI: 10.1016/j.cattod.2007.08.009

Chapter 3

Effect of Silicon Content in Functional Properties of Thin Films

Henry S. Vanegas, Jose E. Alfonso and Jhon J. Olaya

Abstract

Silicon (Si) has been the paradigm of the electronic industry, because it has been used in fabrication of different electronic circuit elements such as diodes, operational amplifiers, transistors, and in the last few decades, it has been the base material of the development of solar energy industry. However, other research fields in which the physical and chemical properties of Si have demonstrated to be relevant to applications in areas such as the mechanical, optical, electrical, and electrochemical are less known. Therefore, it is relevant to know the technology that has generated materials with new or better physical and chemical properties, such as the nanocomposite materials that have been growing in thin films. These films exhibiting new properties with respect to the bulk material and with the addition of Si, have demonstrated to improve the properties of the transition metal nitride (MeN), obtaining thin films with a high nanohardness, wear resistance, corrosion resistance, and high thermal stability. Therefore, the main objective of this chapter is to know the role that plays the incorporation Si in growth, microstructure, chemical composition, and functional properties of ZrN thin films.

Keywords: silicon, thin films, incorporation, mechanical, optical, electrical

1. Introduction

Thin films have been widely used in different application fields such as in chemical as diffusion barrier, in electrochemical as films for protection against corrosion, in mechanical as hard and wear resistance coating, in optical as reflection and antireflection coating, and in electronic area as conductor or insulator material [1–12].

Different chemical elements or compounds have been used to deposit thin films, which have been shown to improve the surface properties of a material (substrate). Among all these compounds, we find that the transition metal nitrides (MeN) have been widely used in mechanical applications due to their high hardness and wear resistance [2, 13–16]. However, it has been found that the addition of a third element may improve the physical and chemical properties of MeN due to a change that this new element generates in the microstructure of these materials [1, 17–30]. These structures are called nanocomposite, and there are two different groups of hard nanocomposite films, for instance: (i) crystalline/amorphous nanocomposites and (ii) crystalline/crystalline nanocomposites.

An element used to improve the properties of MeN is the silicon. In the published literature (**Table 1**), it has been found that the Si addition generated the formation of two phases: a nanocrystalline of MeN and an amorphous phase of silicon nitride (Si_3N_4), which improved the physical and chemical properties of MeN [17, 19–21,

Nanocomposite film	Deposition technique
ZrSiN	Reactive magnetron cosputtering [17, 24, 28, 32, 33, 35, 44, 46, 48, 49, 51–55, 57, 59, 63]
ZrSiN	Unfiltered cathodic arc evaporation [21, 23]
CrAlSiN	Cathodic arc evaporation [66]
TiSiN	Chemical vapor deposition in a fluidized bed reactor at atmospheric pressure (AP/FBR-CVD) [70]
TiSiN	A combination of DC and RF magnetron sputtering [41, 64]
TiSiN	Vacuum cathodic arc evaporation [58]
zAlSiN	DC magnetron sputtering [19, 47, 71]
WSiN	DC reactive unbalanced magnetron sputtering [49, 72]
TiAlSiCuN	DC reactive magnetron sputtering [65]
ZrTiCrNbSiN	Vacuum arc evaporation [31]
CrZrSiN	Unbalanced magnetron sputtering [37]
TiSiN-Ag	Reactive magnetron cosputtering [25]
TaSiN and CrTaSiN	Reactive magnetron cosputtering [36, 50]
AlTiSiN and CrSiN	Cathode arc ion plating system [34, 73]
TiSiCN	Conventional magnetron sputtering and plasma enhanced magnetron sputtering (PEMS) [38, 74]
ZrSiN	Hybrid cathodic arc and chemical vapor process [56]
TiAlVSiN	Vacuum cathodic arc evaporation [58]
NbSiN	Unbalanced magnetron sputtering [71]
CrSiN	Closed field unbalanced magnetron sputtering [75]

Table 1.
Deposition techniques used for depositing nanocomposites thin films.

23–25, 28, 31–66]. It has been demonstrated that these nanocomposite thin films have high nanohardness, wear resistance, corrosion resistance, and thermal stability at high temperature. However, the properties of the films depend on the deposition technique and the growth parameters used [4, 8, 10, 13, 26, 67–69].

Among the papers published on nanocomposite films, one of the researchers who has contributed the most in the development of these materials have been Veprek and their co-workers [64, 76]. They have reported the formation of Ti—Si—N nanocomposites films with a hardness >40 GPa, which consist of a nanocrystalline phase (TiN) embedded on the matrix amorphous (Si_3N_4), produced by sputtering magnetron. They found that the decrease in the crystalline size and the formation of two phases (nanocomposite) improve the hardness films to hinder the multiplication and movement of dislocation and the growth of flaws. However, the hardness cannot improve in small nanocrystals about <10 nm, due to the fraction of the material in the increasing grain boundary generates a decrease in the hardness of the film by a grain boundary sliding (Hall-Petch relationship). Therefore, they suggest that an increase on the strength and hardness of the films can be archived with decreasing crystalline size only if grain boundary sliding can be blocked, and this behavior has been shown in different nanocomposite films. Also, they proposed one model to explain the formation of nanocomposite, explaining the phase segregation in Ti—Si—N because of spinodal decomposition during deposition. The spinodal decomposition process consists of the reduction of the solubility limit of the silicon on MeN lattice, generating the complete phase segregation of the SiN_x around of the MeN crystals or an increase of the thickness of SiN_x amorphous phases [48, 76]. This process is obtained with high temperature (>550°C) deposition and high nitrogen pressure [64].

The scientific and technologic importance of the development of the nanocomposite films with the chemical and physical performances that those has shown, it is evident if it is considered the great amount the publications that over this theme has been produced. In **Table 1**, it is summarized the most important works of nanocomposite thin films with silicon, as can see, the deposition method most used to deposit these films is sputtering technique. This technique allows to deposit metallic and insulator elements and compounds at low temperature, maintaining the composition of the target. In addition, the films deposited have shown to be of high quality (homogeneous and dense) and with good adhesion [3, 77, 78].

In this chapter, we discussed about of a new generation of nanocomposite films composed of two phases: a nanocrystalline embedded in an amorphous matrix, specifically of nanocomposites formed by zirconium nitride and silicon.

Therefore, the content of the chapter is divided into the following topics: Section 1 is dedicated to describe the more usually experimental deposition conditions of the films via magnetron sputtering techniques. In Section 2 of the chapter will present chemical analysis of their bulk through spectroscopy of the X-ray dispersive (EDX) and the chemical surface analysis of the films by means of spectroscopy of photoelectrons (XPS). Section 3 is dedicated to discuss the influence of silicon in the crystalline structure of the films. This analysis is done through X-ray diffraction (XRD) and transmission electron microscopy (TEM). Sections 4–6 will describe the electrical, optical, and mechanical behavior of the deposited films, respectively. Finally, Section 7 will present the corrosion resistance that gives the films on stainless steel substrates. This analysis will be done with potentiodynamic polarization curves (TAFEL).

2. Growth of thin films using DC reactive magnetron sputtering technique

Most publication about nanocomposite films has shown that the DC reactive magnetron cosputtering technique is the most used to deposit these materials (see **Table 1**), and they have reported that the formation and microstructure of the films are determined by the deposition parameters, such as applied power at the target, working pressure, bias voltage, and deposition temperature (see **Table 2**).

In a sputtering process, the surface target is hitting with ions produced by an electric discharge, which form plasma. Normally, no reactive gas (Argon-Ar) is used to form the plasma. The interaction of these ions with the surface of the target causes the atoms on the surface to be ripped off through a moment exchange between ions and atoms of the target [4, 77, 79]. These sputtered atoms must transit from target to the substrate surface. During this displacement, the sputtered atoms experience many collisions with the particles that are in this region (sputtered atoms or Ar atoms or reactive atoms in the case of a reactive gas). These collisions change the velocity, direction, and the energy of the sputtered atoms. Therefore, the number of atoms that reach to the substrate surface will depend on the working pressure and the target-substrate distance. Moreover, the formation of the film is related with the condensing energy of the atoms (adatoms) on the surface of the substrate. Different works have found that amorphous films are formed when the adatoms have low energy of diffuse on the surface that does not allow that they may find low energy sites for the nucleation; while, a crystalline structure may be formed when the adatoms have a high surface mobility [8, 15, 80–88]. However, several works have found that when the growing films are exposed to bombardment

Nanocomposite	T[a] (C)	Target	Gas[b] (sccm)	Power[c] (W)	Power[d] (W)	W.P[e] (Pa)	Bias (V)
ZrSiN [17, 33, 39]	RT 200 500 900	Zr (1) Si (2) Zr + Si pellets	Ar:19 N_2:2 Ar + N_2 Ar:8 N_2:4	Zr: 120–150[f]	Si: 0–250[g]	4×10^{-1} 2.7×10^{-1}	0–100
TiSiN [41]	200	Ti (1) Si (2)	Ar + N_2	Ti: 300–500[f]	Si: 50–150[g]	0.9–1.2	25
AlSiN [19, 47]	400 500	Al-Si alloy Al-Si	Ar + N_2	75	0	1	NM
WSiN [49, 72]	500	WSi_2 alloy	Ar + N_2	NM	0	0.05–0.5	100
TaSiN [50]	500	$TaSi_2$ alloy	Ar + N_2	NM	0	5×10^{-1}	50
CrSiN [75]	NM	Cr (1) Si (2)	Ar + N_2	Cr: 20 cm^{-2}	NM	NM	100
TiAlSiCuN [65]	RT	TiAlSi alloy Cu (2)	Ar + N_2	TiAlSi: 537.5	Cu: 0–21.25	6×10^{-1}	NM
CrZrSiN [37]	120	CrZrSi segment	Ar + N_2	0.7 k[g]	0	5×10^{-1}	NM
CrTaSiN [36]	NM	Cr (1) Ta (2) Si (3)	Ar: 12 N2: 8	Cr: 150[g]	Ta: 100[f] Si: 150[f]	4×10^{-1}	NM

RT and NM are used to room temperature and when the variable is not mentioned, respectively.
[a]*Temperature.*
[b]*Gas type and flux.*
[c]*First target power.*
[d]*Second target power.*
[e]*Working pressure*
[f]*Direct current.*
[g]*Radio frequency.*

Table 2.
Deposition parameters for different nanocomposite films obtained by reactive magnetron sputtering.

of high-energy particle, the structure and properties of deposited films can be improved. This can be achieved by applying a negative voltage called bias to the surface substrate [18, 81, 89]. These high-energy particles can improve the diffusion of the adatoms on the substrate surface. Another way to improve adatoms mobility is by increasing the temperature of the substrate [15, 83, 85]. Therefore, in the deposition of thin films, it is important to be able to find the deposition parameters to determine and understand their physical and chemical properties.

Finally, the composition of the target is another parameter that will affect the characteristics of the films. It has been shown that the addition of silicon to transition metal nitride can affect the structure and properties of the films; due to that the addition of the silicon atoms can block the surface mobility of the adatoms of the metal or the transition metal nitride [28, 51, 57, 63, 90]. Therefore, as the amount of silicon is increased, the structure of the film is changed from a polycrystalline, nanocrystalline to amorphous.

Zirconium nitride with silicon films was deposited via DC reactive magnetron sputtering technique. Only one zirconium target was used with silicon pellets on the target surface to tailor the Si content in the Zr—Si—N films. The deposition parameters used are shown in **Table 3**. This sputtering method has been used for depositing different nanocomposite films such as: Ti—Si—N [91], W—Si—N [72], Zr—Si—N [92], Al—Si—N [93], and Nb—Si—N [71].

Deposition parameters	Value
Target of sputter	Zr (99.99%)
Reactive gas	Nitrogen
Target diameter (cm)	5
Distance target-substrate (cm)	5
Base pressure (Pa)	4×10^{-4}
Working pressure (Pa)	8×10^{-1}
Temperature (°C)	200
Voltage of bias (V)	0
Applied power (W)	140
Number of Si pellets	0, 1, 2
Deposition time (s)	3600

The Ar/N_2 flow ratio was optimized to obtain ZrN films.

Table 3.
Deposition parameters used to deposit ZrN-Si films via DC reactive magnetron sputtering.

The zirconium nitride and silicon nitride were obtained by adding of nitrogen as reactive gas to the deposition chamber. Three series of samples were deposited with different numbers of silicon pellets (X), ZrN + XSi, where X = 0, 1, and 2. Finally, thicknesses of the deposited film were obtained from the cross-sectional scanning electronic microscope images, which are not included in this chapter. These images showed that the thicknesses of the films were 823±2, 955 ±8, and 1060 ±6 nm for ZrN, ZrN + 1Si, and ZrN + 2Si, respectively. The results of the chemical and structural characterization and the study of the functional properties are shown in the following subsections.

3. Chemical characterization by means of spectroscopy of the X-ray dispersive (EDX) and spectroscopy of photoelectrons (XPS)

In the sputtering technique, the elements of target are transferred to the substrate surface; this is verified with EDX analysis. **Figure 1a** shows the EDX spectrum for the ZrN + 2Si film.

The EDX spectrum evidences the presence of zirconium, nitrogen, and silicon in the film. The elemental chemical composition of the deposited films is listed in **Table 4**. **Figure 1b** shows the variation of the zirconium content with the increase of the silicon content in the films. As silicon content increases, the Zr content decreases in the films due to the reduction of the effective sputtering area of the Zr target with the Si pellets. These results are similar to those published by other authors using the same sputtering configuration [59, 62, 92]. The EDX results also showed that with one Si pellet, the Si content was of 8 at.% and with two pellets was of 15 at.%.

It has been found that when the solubility limit of Si in MeN lattice is exceeded, the Si atoms form a Si_3N_4 phase [94]. The formation of Si_3N_4 phase into MeN grain boundaries is typical for the Me—Si—N systems [17, 19, 25, 41, 71, 73]. Therefore, in our case, a chemical analysis for XPS of the ZrN—Si deposited films with different Si contents was carried out to show the formation of the Si_3N_4 phase with the Si addition. **Figure 2** shows the high-resolution XPS spectrum for the MeN films. The XPS results of Zr 3d peaks (**Figure 2a**) showed the presence of Zr—N bond with a binding energy of 179.6 eV [95], and the Si 2p peaks (**Figure 2b**) showed the presence of Si-N bond to 100.8 eV [24] on the film surface. Additionally, the results

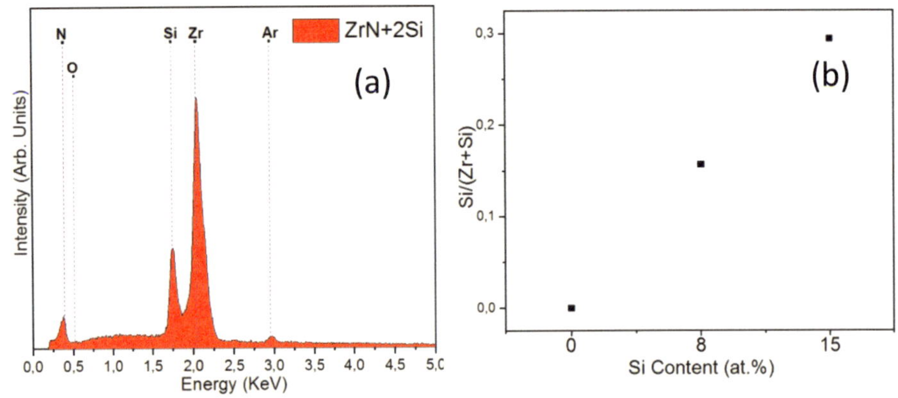

Figure 1.
(a) EDX spectrum of ZrN + 2Si film with a 15 at.% of silicon and (b) the stoichiometry behavior of the ZrN—Si deposited film at different silicon contents.

Samples name	Atomic percentage (at.%)		
	Zirconium (Zr)	Nitrogen (N)	Silicon (Si)
ZrN	51.0	49.0	0.0
ZrN + 1Si	43.0	49.0	8.0
ZrN + 2Si	36.0	49.0	15.0

Table 4.
The elemental chemical composition of the deposited films with different Si contents.

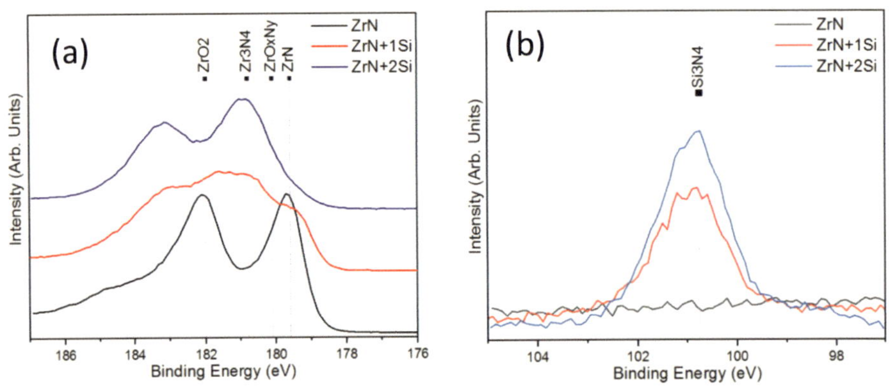

Figure 2.
High-resolution XPS spectra for ZrN—Si films with different silicon contents (a) Zr 3d and (b) Si 2p. The Si addition generated the formation of the Si_3N_4 phase in the films.

showed the formation of zirconium oxide (ZrO_2) and oxynitride of zirconium (ZrO_xN_y) that can be due to the presence of residual oxygen in the deposition chamber and to high possibility that has the zirconium to react with oxygen according to enthalpy of formation for ZrO_2 that is −1101.3 kJ/mol [96].

Various works have shown that the Si exists as solid solution in the ZrN lattice up to 3.0 at.%, but when the Si content increases, the formation of the Si_3N_4 phase is observed [17, 21]. Therefore, the EDX and XPS results show that with a Si content of 8 at.%, the solubility limit of Si in ZrN lattice is exceeded, generating the formation of Si_3N_4 into ZrN grain boundaries with the increased Si content, probability of the volume of the phase of Si_3N_4 is increased, and the phase of ZrN is decreased.

4. Structural characterization of the films through X-ray diffraction (XRD) and transmission electron microscopy (TEM)

With the addition of Si, it has been found that the microstructure of MeN films changes, and this change will depend on Si. Three types of microstructure have been observed in function of Si content: polycrystalline films with low Si content up to 3 at.%, nanocrystalline films with 3–10 at.% (nanocomposite: nanocrystalline and amorphous phase), and with a Si content above 10 at.%, the films are amorphous. These values on the silicon content are obtained for Ti—Si—N deposited films with different Si contents [97], but may change depending on MeN. In our case, the microstructure ZrN—Si deposited films were characterized by XRD and TEM techniques. **Figure 3** shows the XRD pattern of ZrN films with different Si contents deposited on the common glass substrate. **Figure 3** exhibits diffraction peaks corresponding to fcc-ZrN (pdf. 01-078-1420) for the ZrN film without silicon (black color). The addition of Si, red and blue color, indicates that the diffraction peak of the ZrN (111) tends to broaden, while the ZrN (200) peak disappears as Si content increased. The broadening of the peak may be due to the formation of nanocrystals of cubic ZrN and tetragonal ZrO_2, reported in the 2θ position 33.83° (pdf. 01-078-1420) and 30.27° (pdf. 00-050-1089), respectively. The crystalline size for ZrN films is <10 nm, which was determined for the Scherrer equation, and with the addition of Si, the crystalline size decreased until 5 nm. The XRD evidenced that Si addition generated a refinement of grain, which is related with a broadening of the diffraction peaks. With a high Si content (15 at.%), the film is amorphous.

To study the structure of the ZrN—Si film with Si content of 8 at.% in more detail, transmission electron microscopy with selected area electron diffraction (SAED) was done. **Figure 4** shows the SAED pattern of ZrN + 1Si film. It shows the presence of the (111), (200), and (220) diffraction rings, which indicate a fcc-ZrN structure, but the (111) diffraction ring is very broad, which is in very good agreement with the XRD results in the same d-spacing from 0.295 to 0.262 nm.

In addition, this ring broadens may be related with a mixture of phases, such as: ZrN, ZrO_2 and Si_3N_4 as we can see in **Figure 5**. This figure shows the XRD pattern of ZrN + 1Si film and the crystallographic databases for ZrN (pdf. 01-078-1420), ZrO_2 (pdf. 00-050-1089) and Si_3N_4 (pdf. 00-033-1160).

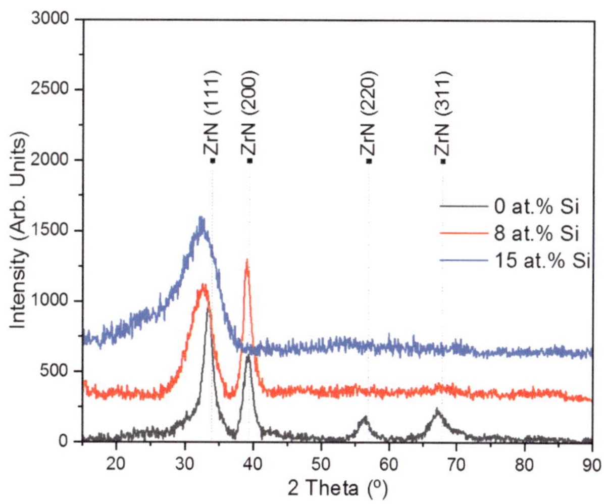

Figure 3.
The XRD patterns of the ZrN—Si films with different silicon contents.

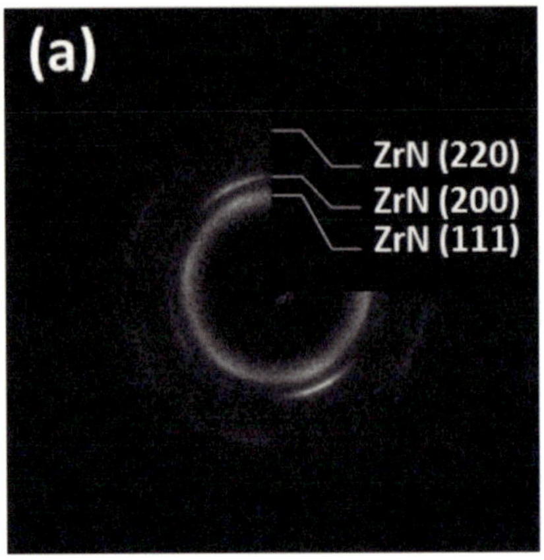

Figure 4.
(a) SAED pattern of ZrN + 1Si film. The diffraction ring, from 0.295 to 0.262 nm, is diffused possibly by the presence of various crystalline phases.

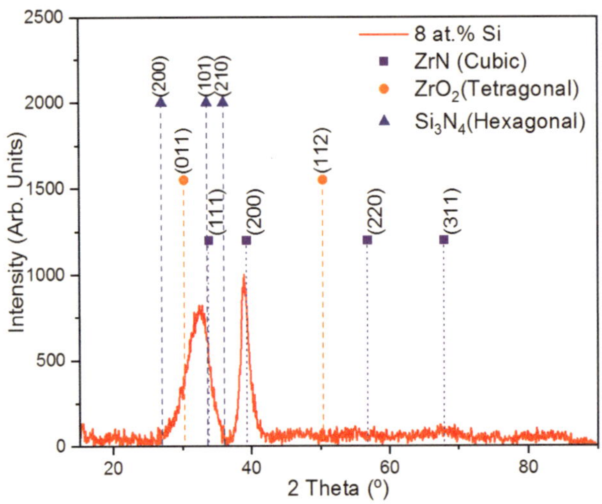

Figure 5.
The XRD pattern of the ZrN + 1Si film with crystallographic databases for ZrN (cubic), ZrO_2 (tetragonal), and Si_3N_4 (hexagonal).

The different crystallographic phases present in the ZrN + 1Si film were analyzed by phase contrast images, **Figure 6**. The existence of nanocrystals and the Fast Fourier Transform (FTT) in **Figure 6b** confirms the presence of the diffraction rings observed in **Figure 4**.

However, this figure allows separating two rings at the lowest d-spacing, which may confirm the hypothesis of a mixture of phases, as observed with XPS and XRD results. According to ZrN (01-078-1420) and ZrO_2 (00-050-1089) pattern diffraction files, the interplanar distances in **Figure 6b** correspond to the ZrN face center cubic and ZrO_2 tetragonal. Finally, the XPS, XRD, and HRTEM results show that with a Si content 0 at.%, the ZrN film is polycrystalline, with a 8 at.%, the film is nanocrystalline (ZrN and ZrO_2 nanocrystalline), and possibly, with a Si_3N_4 amorphous matrix and with a 15 at.%, the Zr—Si—N film is amorphous.

Various published literatures have reported that electrical, optical, mechanical, and electrochemical properties of the nanocomposite films depend on their nanostructure. These works have found that with the addition of silicon to binary MeN,

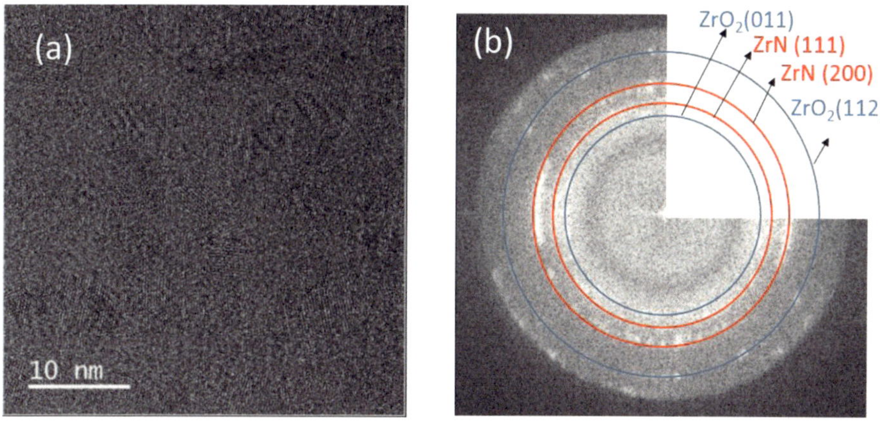

Figure 6.
(a) HR-TEM image of ZrN + 1Si film and (b) SAED pattern of the image (a). The results allow to identify the presence of two different crystalline phases between 0.295 and 0.262 nm.

hardness, thermal stability, and corrosion resistance of the films can be improved. In order to illustrate the relationship between the microstructure and functional properties, measurements of resistivity, reflectance, transmittance, nanohardness, and potentiodynamic polarization were carried out.

5. Electrical properties

Electrical resistivity and sheet resistance measurements were obtained through the van der Pauw method for the Zr—Si—N deposited films, and their values were calculated and listed in **Table 5**. The results evidence that the electrical resistivity increases from 4.40×10^{-4} Ω cm (free Si) to 77.99 Ω cm (with 8 at.% Si) with the addition of silicon. This increase on the resistivity has been reported by other authors in different nanocomposites [54, 98]. They have found that depending on the chemical composition and electrical nature of the amorphous phase and nanocrystalline phase, the resistivity of Me—Si—N nanocomposite films can change. The electrical resistivity increases with increasing Si content, and the nanocomposite films have showed to have a structure of MeN nanocrystalline (conductor) surrounded of a SiN_x amorphous phase (insulator). However, when the electrical resistivity behavior is independent to Si content, the resistivity is due to a direct percolation of the MeN_{1-x} nanocrystalllines (conductors) separated by low degree of nitration of the SiN_x grain boundary phase [98].

Therefore, the results obtained evidenced the formation of Zr—Si—N nanocomposite films with ZrN nanocrystallites embedded in the amorphous phase of SiN_x, and the increase in the electrical resistivity with the Si addition is due to an increase in the thickness of SiN_x layer that covers the nanocrystallites. The grain boundary scattering model is used for explaining the electrical conductivity in nanocomposite films [98].

Film	Silicon (at.%)	Resistivity (Ω cm)	Sheet resistance (Ω/■)
ZrN	0	4.40×10^{-4}	5.35
ZrN + 1Si	8	77.99	817006.68

The sample with a 15 at.% Si was not possible to measure the electrical resistivity due to high resistivity.

Table 5.
Values of resistivity the ZrN with different Si contents.

6. Optical properties

The optical properties were investigated using a UV–Vis–NIR spectrophotometer. Reflectance and transmittance measurement were carried out from 300 to 2500 nm. The reflectance spectra of ZrN (black line) and ZrN + 1Si (red line) films are shown in **Figure 7**. In this figure, the typical reflectance of ZrN is observed [85]. It exhibits a maximum reflectance in the infrared region, which decreases as wavelengths decrease, and for wavelengths <500 nm, the reflectance slightly increases again. The ZrN films exhibit a similar Drude-like behavior [53, 85]. At longer wavelengths, the high electromagnetic absorption in this optical region is due to the conduction electrons and to the absorption at shorter wavelengths is due to the inter-band transitions of the bounded electron [85]. However, with the addition of silicon, the reflectance decreases drastically. At longer wavelengths, the film without silicon has a reflectance <80%, but the film with a Si content of 8 at.% has a reflectance <20%.

Therefore, transmittance measurements were done to investigate the effect of silicon in the ZrN films. The transmittance spectra of ZrN + 1Si and ZrN + 2Si films are shown in **Figure 8**.

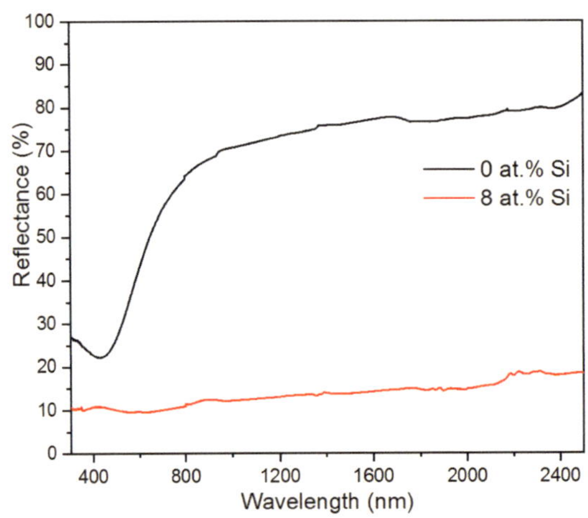

Figure 7.
Reflectance spectra of ZrN and ZrN + 1Si films from 300 to 2500 nm. As Si content increases, the transmittance optical of the films also increase.

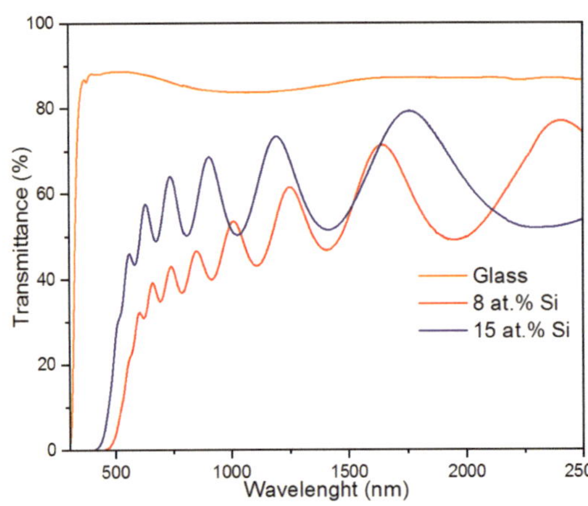

Figure 8.
Transmittance spectra of ZrN—Si films with different Si contents and of the bare substrate (common glass). As the silicon content increases, the transmittance of the deposited films increases.

The transmittance spectra show that as silicon content increases, the transmittance also increases. With the addition of Si, new phases are generated according to previous results (TEM and XPS), such as: SiN_x, ZrN, and ZrO_2 phases. These phase mixtures change the optical behavior of the deposited films due to SiN_x and ZrO_2 that have been reported as optically transparent materials, while ZrN is an optically reflective material. Nowadays, there is not a theoretical model that may explain this behavior. However, the increase in the transmittance in films may be due to an increase in the volume of SiN_x phase and a decrease of ZrN nanocrystalline phase with the increase of Si content.

7. Mechanical properties

The nanohardness (H) values as a function of Si content for the deposited films are shown in **Figure 9** and **Table 6**. The H for the ZrN film is 29.55 ± 3.70 GPa, which decreases with the addition of silicon to 18.12 ± 2.65 and 15.92 ± 1.23 GPa at 8 and 15 at.% of Si, respectively. The value obtained for ZrN was similar to the reports from other authors [99, 100]. The decrease of the nanohardness with the Si addition is related with the increase amorphous phase of SiN_x and decreasing of the crystalline size of ZrN [101].

The mechanical properties of nanocomposite films depend on the chemical composition of each one of the phases present, of the crystallite size, crystallographic orientation, lattice structure, and the thickness grain boundary phase [94, 101]. Different works have reported that the main mechanisms that allow to explain the hardness enhancement in the nanocomposites are three: (i) the dislocation-induced plastic deformation when the crystalline size is >10 nm, (ii) the nanostructure of materials when the crystalline size is ≤10 nm, and (iii) cohesive force between atoms when the crystalline size is <10 nm. However, when the thickness of amorphous phase is larger than the crystalline size, the nanohardness of the films decreases due to a deformation mechanism reported as grain boundary sliding [24, 94].

According to the XPS and electrical resistivity value results, the Si addition generated the formation of an amorphous phase of SiN_x, which increases its thickness with the silicon content in the film. It has been reported that when

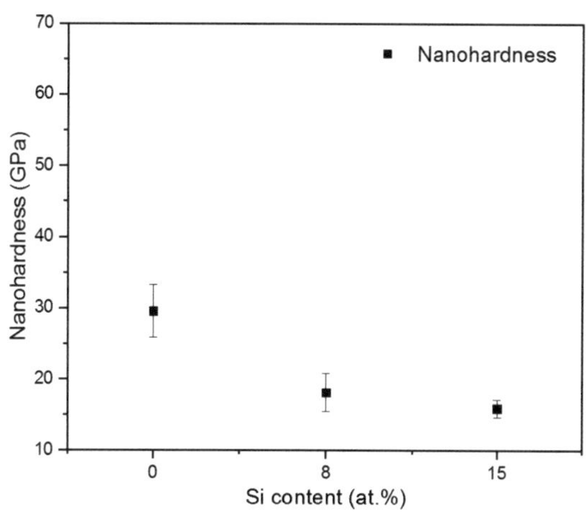

Figure 9.
Nanohardness of the films as a function of silicon content. As Si content increases, the nanohardness of films decreases.

Samples	Silicon (at.%)	Nanohardness (H) GPa
ZrN	0	29.55 ± 3.70
ZrN + 1Si	8	18.12 ± 2.65
ZrN + 2Si	15	15.92 ± 1.23

Table 6.
Results from nanohardness tests.

the SiN_x phase thickness is larger than the crystallite size of the ZrN phase, the nanohardness of the films decreases due to an increase of the volume fraction of the amorphous soft phase. The deformation mechanism, in this case, is grain boundary sliding [24].

8. Electrochemical properties

Potentiodynamic polarization curves of the 316l stainless steel (substrate) and deposited film with different Si contents were carried out for studying the Si effect in electrochemical properties of the films. They were tested in 3.5 wt.% NaCl solution, and the results are shown in **Figure 10**. For each curve, corrosion potential (E_{corr}) and corrosion current density (J_{corr}) were determined and are reported in **Table 7**. The results show that the deposited film with a Si content of 15 at.% has lower J_{corr} than that of the uncoated SS 316 L substrate, indicating that with the Si addition or with the coating, the corrosion resistance increases.

Several research groups have found that the addition of silicon to MeN films can improve the corrosion resistance due to the formation of nanocomposite films. The increase on the corrosion resistance could be attributed to the formation of a dense structure, which can block the paths of corrosion medium to the substrate. It has been demonstrated to nanocomposite films, such as: Ti—Si—N [102], Al—Si—N [103], Nb—Si—N [71], and W—Si—N and Zr—Si—N [104].

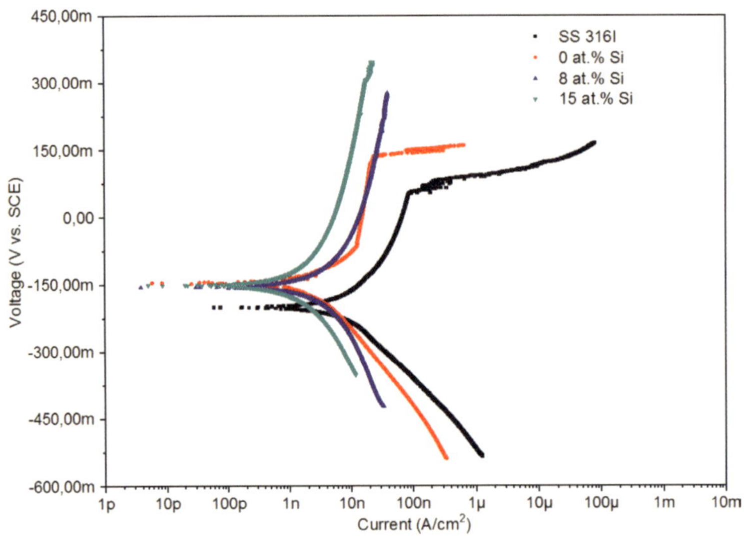

Figure 10.
Potentiodynamic polarization curves for the films and substrate SS 316L.

Sample	Silicon (at.%)	J_{corr} (nA/cm^2)	E_{corr} (mV)
SS 316l	0	7.13 ±9	−198 ± 94
ZrN	0	3.19 ± 7	−146 ± 69
ZrN + 1Si	8	4.83 ± 4	−155 ± 47
ZrN + 2Si	15	2.28 ± 9	−150 ± 93

Table 7.
Results from potentiodynamic polarization curves, corrosion current density (I_{corr}), and corrosion potential (E_{corr}) for each sample.

9. Conclusions

The addition of a third element as silicon affects the microstructure and functional properties of the transition metal nitride films, specifically those of zirconium nitride as observed in this chapter. Therefore, it is important has a control of the content of this element in the deposited film to understand and relate the microstructure and their properties.

With a silicon content of 8 at.% and with the deposition parameter used, the microstructure changed from a polycrystalline structure of fcc-ZrN (free silicon) to a mixture of nanocrystalline (ZrN) and amorphous (SiN_x) phases, and with an increase in the Si content (15 at.%), the films were amorphous. However, these films showed the formation of two crystalline phases corresponding to zirconium nitride and zirconium oxide due to the base pressure used, which is not high enough to remove oxygen in the deposited chamber, and high enthalpy of formation for ZrO_2.

These changes in the microstructure and the mixture of phases present in the films generated changes in the functional properties of zirconium nitride:

- With the addition of silicon, the electrical resistivity increased various orders of magnitude comparison with the resistivity of ZrN. The electrical measurements allowed to determine that the films have a nanocomposite structure: nanocrystalline of ZrN conducting embedded in an amorphous SiN_x insulating phases.

- As silicon content increases, the optical response changed from a high reflectance in the infrared region (ZrN) to a high transmittance in the infrared region (ZrN—Si coatings).

- The nanohardness values decrease from 29.55 GPa (free silicon) to 15.92 GPa (15.0 at.% Si), due to an increase in the thickness of amorphous phase (SiN_x) and a decrease in crystalline size (ZrN).

- The potentiodynamic polarization curves showed that the coated substrate has higher corrosion resistance than the uncoated substrate due to a decrease in I_{corr}.

Acknowledgements

The authors are grateful to Professor Sandra Carvalho of the Minho University (Portugal) who performed the EDS and XPS measurements and the professors Sebastian Calderon and Paulo Ferreira of INL-International Iberian Nanotechnology Laboratory who performed HRTEM.

The authors are also grateful to the Universidad Nacional de Colombia (UNAL) for its financial support through the projects: “Recubrimientos Funcionales de (Zr, Ag, Si)N y (Zr, Cu, Si)N producidos por la técnica de Co-Sputtering Magnetrón reactive” and “Resistencia a la corrosión de recubrimientos de (Ag, Zr, Si)N y (Ti, Zr, Si)N producidos por Co-Sputtering Reactivo”.

Author details

Henry S. Vanegas[1], Jose E. Alfonso[2]* and Jhon J. Olaya[1]

1 Mechanics and Mechatronics Department, Universidad Nacional de Colombia, Bogotá, Colombia

2 Department Physics, Universidad Nacional de Colombia, Bogotá, Colombia

*Address all correspondence to: jealfonsoo@unal.edu.co

References

[1] AL-Rjoub A, Rebouta L, Costa P, Vieira LG, Miranda TMR, Barradas NP, et al. CrAlSiN barrier layer to improve the thermal stability of W/CrAlSiN_x/CrAlSi$O_y$$N_x$/SiAl$O_x$ solar thermal absorber. Solar Energy Materials & Solar Cells. 2019;**191**:235-242. DOI: 10.1016/j.solmat.2018.11.023

[2] Gharavi MA, Greczynski G, Eriksson F, Lu J, Balke B, Fournier D, et al. Synthesis and characterization of single-phase epitaxial Cr2N thin films by reactive magnetron sputtering. Journal of Materials Science. 2019;**54**(2):1434-1442. DOI: 10.1007/s10853-018-2914-z

[3] Simon AH. Sputter processing. In: Seshan K, editors. Handbook of Thin Film Deposition. 3rd ed. William Andrew; 2012. p. 55-88. DOI: 10.1016/C2009-0-64359-2

[4] Krishna SB. In: Babu Krishna Moorthy S, editor. Thin Film Structures in Energy Applications. London: Springer International Publishing; 2015. DOI: 10.1007/978-3-319-14774-1

[5] Snihirova D, Lamaka SV, Montemor MF. Smart composite coatings for corrosion protection of aluminium alloys in aerospace applications. In: Montemor MF editor. Smart Composite Coatings and Membranes. Woodhead; 2016. p. 85-121. DOI: 10.1016/B978-1-78242-283-9.00004-X

[6] Gao B, Li X, Ding K, Huang C, Li Q, Chu PK, et al. Recent progress in nanostructured transition metal nitrides for advanced electrochemical energy storage. Journal of Materials Chemistry A. 2019;7(1):14-37. DOI: 10.1039/C8TA05760E

[7] Beake BD, Harris AJ. Nanomechanics to 1000 C for high temperature mechanical properties of bulk materials and hard coatings. Vacuum. 2019;**159**:17-28. DOI: 10.1016/j.vacuum.2018.10.011

[8] Frey H, Khan HR. Handbook of Thin-Film Technology. Berlin-Heidelberg: Springer; 2015. 379 p. DOI: 10.1007/978-3-642-05430-3

[9] Kim M, Kang T-W, Kim SH, Jung EH, Park HH, Seo J, et al. Antireflective, self-cleaning and protective film by continuous sputtering of a plasma polymer on inorganic multilayer for perovskite solar cells application. Solar Energy Materials & Solar Cells. 2019;**191**:55-61. DOI: 10.1016/j.solmat.2018.10.020

[10] Zhu H, Dong Z, Niu X, Li J, Shen K, Mai Y, et al. DC and RF sputtered molybdenum electrodes for Cu(In,Ga)Se_2 thin film solar cells. Applied Surface Science. 2019;**465**:48-55. DOI: 10.1016/j.apsusc.2018.09.130

[11] Yue C, Jiang S, Zhu H, Chen L, Sun Q, Zhang D. Device applications of synthetic topological insulator nanostructures. Electronics. 2018;7(10):225. DOI: 10.3390/electronics7100225

[12] Albella JM. Depósito mediante pulverización catódica (sputtering). In: Albella J, editor. Láminas Delgadas y Recubrimientos: Preparación, Propiedades y Aplicaciones. Madrid: Consejo Superio de Investigaciones Cientificas; 2003. pp. 147-167

[13] Bouaouina B, Besnard A, Abaidia SE, Airoudj A, Bensouici F. Correlation between mechanical and microstructural properties of molybdenum nitride thin films deposited on silicon by reactive R.F. magnetron discharge. Surface and Coatings Technology. 2018;**333**:32-38. DOI: 10.1016/j.surfcoat.2017.10.028

[14] Haubner R, Lessiak M, Pitonak R, Köpf A, Weissenbacher R. Evolution of conventional hard coatings for its use on cutting tools. International Journal of

Refractory Metals and Hard Materials. 2017;**62**:210-218. DOI: 10.1016/j.ijrmhm.2016.05.009

[15] Kumar DD, Kumar N, Kalaiselvam S, Dash S, Jayavel R. Wear resistant super-hard multilayer transition metal-nitride coatings. Surfaces and Interfaces. 2017;**7**:74-82. DOI: 10.1016/j.surfin.2017.03.00

[16] Singh A, Kuppusami P, Khan S, Sudha C, Thirumurugesan R, Ramaseshan R, et al. Influence of nitrogen flow rate on microstructural and nanomechanical properties of Zr-N thin films prepared by pulsed DC magnetron sputtering. Applied Surface Science. 2013;**280**:117-123. DOI: 10.1016/j.apsusc.2013.04.107

[17] Silva Neto PC, Freitas FGR, Fernandez DAR, Carvalho RG, Felix LC, Terto AR, et al. Investigation of microstructure and properties of magnetron sputtered Zr-Si-N thin films with different Si content. Surface and Coatings Technology. 2018;**353**:355-363. DOI: 10.1016/j.surfcoat.2018.07.106

[18] Dubey P, Martinez G, Srivastava S, Chandra R, Ramana CV. Effect of bias induced microstructure on the mechanical properties of nanocrystalline zirconium tungsten nitride coatings. Surface and Coatings Technology. 2017;**313**:121-128. DOI: 10.1016/j.surfcoat.2017.01.067

[19] Mishra SK, Kumari S. Development of hard and optically transparent Al-Si-N nanocomposite coatings. Surface and Interface Analysis. 2017;**49**(4): 345-348. DOI: 10.1002/sia.5954

[20] Chen Y-I, Lin K-Y, Wang H-H, Cheng Y-R. Characterization of Ta-Si-N coatings prepared using direct current magnetron co-sputtering. Applied Surface Science. 2014;**305**:805-816. DOI: 10.1016/j.apsusc.2014.04.011

[21] Yalamanchili K, Forsén R, Jiménez-piqué E, Jöesaar MPJ, Roa JJ, Ghafoor N, et al. Structure, deformation and fracture of arc evaporated Zr-Si-N hard films. Surface and Coating Technology. 2014;**258**:1100-1107. DOI: 10.1016/j.surfcoat.2014.07.024

[22] Pang Y, Xia C, Wang M, Li Z, Xiao Z, Wei H, et al. Effects of Zr and (Ni, Si) additions on properties and microstructure of Cu-Cr alloy. Journal of Alloys and Compounds. 2014;**582**: 786-792. DOI: 10.1016/j.jallcom.2013.08.146

[23] Warcholinski B, Kuznetsova TA, Gilewicz A, Zubar TI, Lapitskaya VA, Chizhik SA, et al. Structural and mechanical properties of Zr-Si-N coatings deposited by arc evaporation at different substrate bias voltages. Journal of Materials Engineering and Performance. 2018;**27**(8):3940-3950. DOI: 10.1007/s11665-018-3483-7

[24] Chang L-C, Zheng Y-Z, Chen Y-I, Chang S-C, Liu B-W. Bonding characteristics and chemical inertness of Zr-Si-N coatings with a high Si content in glass molding. Coatings. 2018;**8**(5):181. DOI: 10.3390/coatings8050181

[25] Dang C, Li J, Wang Y, Yang Y, Wang Y, Chen J. Influence of Ag contents on structure and tribological properties of TiSiN-Ag nanocomposite coatings on Ti-6Al-4V. Applied Surface Science. 2017;**394**:613-624. DOI: 10.1016/j.apsusc.2016.10.126

[26] Kumar M, Mitra R. Effect of substrate temperature and annealing on structure, stress and properties of reactively co-sputtered Ni-TiN nanocomposite thin films. Thin Solid Films. 2017;**624**:70-82. DOI: 10.1016/j.tsf.2017.01.024

[27] Zheng Y, Li X, Liu Y, Sun W, Dong C. Preparation and characterization of CuN-based ternary alloy films using Cr or Zr for stabilizing N. Journal of Materials Research. 2017;**32**(7):1333-1342. DOI: 10.1557/jmr.2017.62

[28] Chen YI, Chang SC, Chang LC. Oxidation resistance and mechanical properties of Zr-Si-N coatings with cyclic gradient concentration. Surface and Coatings Technology. 2017;**320**: 168-173. DOI: 10.1016/j. surfcoat.2017.01.063

[29] Su J, Yin L, Qin L, Ma N, Huang J. Preparation and performance of ZrAlN anti-reflective coatings for low-emissivity glasses. Ceramics International. 2017;**43**(17):14616-14622. DOI: 10.1016/j.ceramint.2017.07.105

[30] Ren P, Zhang K, He X, Du S, Yang X, An T, et al. Toughness enhancement and tribochemistry of the Nb-Ag-N films actuated by solute Ag. Acta Materialia. 2017;**137**:1-11. DOI: 10.1016/j.actamat.2017.07.034

[31] Pogrebnjak AD, Bagdasaryan AA, Beresnev VM, Nyemchenko US, Ivashchenko VI, Kravchenko YO, et al. The effects of Cr and Si additions and deposition conditions on the structure and properties of the (Zr-Ti-Nb) N coatings. Ceramics International. 2017;**43**(1):771-782. DOI: 10.1016/j. ceramint.2016.10.008

[32] Freitas FGR, Hübler R, Soares G, Conceição AGS, Vitória ER, Carvalho RG, et al. Structural and mechanical properties of Zr-Si-N thin films prepared by reactive magnetron sputtering. Materials Research. 2015;**18**(Suppl 2):30-34. DOI: 10.1590/1516-1439.336214

[33] Ghafoor N, Petrov I, Klenov DO, Freitag B, Jensen J, Greene JE, et al. Self-organized anisotropic $(Zr_{1-x}Si_x)N_y$ nanocomposites grown by reactive sputter deposition. Acta Materialia. 2015;**82**:179-189. DOI: 10.1016/j. actamat.2014.09.029

[34] Yang B, Tian CX, Wan Q, Yan SJ, Liu HD, Wang RY, et al. Synthesis and characterization of AlTiSiN/CrSiN multilayer coatings by cathodic arc ion-plating. Applied Surface Science. 2014;**314**:581-585. DOI: 10.1016/j. apsusc.2014.05.166

[35] Choi H, Jang J, Zhang T, Kim J-H, Park I-W, Kim KH. Effect of Si addition on the microstructure, mechanical properties and tribological properties of Zr-Si-N nanocomposite coatings deposited by a hybrid coating system. Surface and Coatings Technology. 2014;**259**:707-713. DOI: 10.1016/j. surfcoat.2014.10.008

[36] Chen Y, Lin K-Y, Wang H, Lin K. Thermal stability of TaN, CrTaN, TaSiN, and CrTaSiN hard coatings in oxygen-containing atmospheres. Surface and Coatings Technology. 2014;**259**:159-166. DOI: 10.1016/j.surfcoat.2014.02.005

[37] Kim D-J, La J, Kim K-S, Kim S, Lee S. Tribological properties of CrZr-Si-N films synthesized using Cr-Zr-Si segment targets. Surface and Coatings Technology. 2014;**259**:71-76. DOI: 10.1016/j.surfcoat.2014.06.019

[38] El-rahman AMA, Wei R. A comparative study of conventional magnetron sputter deposited and plasma enhanced magnetron sputter deposited Ti-Si-C-N nanocomposite coatings. Surface and Coating Technology. 2014;**241**:74-79. DOI: 10.1016/j.surfcoat.2013.08.049

[39] Cui X, Jin G, Hao J, Li J, Guo T. The influences of Si content on biocompatibility and corrosion resistance of Zr-Si-N fi lms. Surface and Coating Technology. 2013;**228**:524-528. DOI: 10.1016/j.surfcoat.2012.04.060

[40] Bushroa AR, Rahbari RG, Masjuki HH, Muhamad MR. Approximation of crystallite size and microstrain via XRD line broadening analysis in TiSiN thin films. Vacuum. 2012;**86**(8):1107-1112. DOI: 10.1016/j.vacuum.2011.10.011

[41] Bushroa AR, Masjuki HH, Muhamad MR, Beake BD. Optimized

scratch adhesion for TiSiN coatings deposited by a combination of DC and RF sputtering. Surface and Coatings Technology. 2011;**206**(7):1837-1844. DOI: 10.1016/j.surfcoat.2011.07.048

[42] Sheng SHH, Zhang RFF, Veprek S. Phase stabilities and decomposition mechanism in the Zr-Si-N system studied by combined ab initio DFT and thermodynamic calculation. Acta Materialia. 2011;**59**(1):297-307. DOI: 10.1016/j.actamat.2010.09.033

[43] Pogrebnyak AD, Sobol' OV, Beresnev VM, Turbin PV, Dub SN, Kirik GV, et al. Features of the structural state and mechanical properties of ZrN and Zr(Ti)-Si-N coatings obtained by ion-plasma deposition technique. Technical Physics Letters. 2009;**35**(10):925-928

[44] Zhang X, Byrne MS, Lad RJ. Structure and optical properties of $Zr_{1-x}Si_xN$ thin films on sapphire. Thin Solid Films. 2009;**518**(5):1522-1526. DOI: 10.1016/j.tsf.2009.09.025

[45] Jianfeng W, Ma D, Zhongxiao S, Wu T, Kewei X. Study on microstructure and properties of Zr-Si-N films with different nitrogen partial pressures. Rare Metals Materials And Engineering. 2009;**38**(5):753-756

[46] Sandu CS, Cusnir N, Oezer D, Sanjinés R, Patscheider J. Influence of bias voltage on the microstructure and physical properties of magnetron sputtered Zr-Si-N nanocomposite thin films. Surface and Coatings Technology. 2009;**204**(6-7):969-972. DOI: 10.1016/j.surfcoat.2009.06.042

[47] Musil JŠM, Zeman PČR, He D, Han JG, Satava V. Properties of magnetron sputtered Al-Si-N thin films with a low and high Si content. Surface and Coating Technology. 2008;**202**:3485-3493. DOI: 10.1016/j.surfcoat.2007.12.024

[48] Sandu CS, Sanjinés R, Medjani F. Control of morphology (ZrN crystallite size and SiN_x layer thickness) in Zr-Si-N nanocomposite thin films. Surface and Coating Technology. 2008;**202**:2278-2281. DOI: 10.1016/j.surfcoat.2007.09.003

[49] Zeman P, Musil J. Difference in high-temperature oxidation resistance of amorphous Zr-Si-N and W-Si-N films with a high Si content. Applied Surface Science. 2006;**252**:8319-8325. DOI: 10.1016/j.apsusc.2005.11.038

[50] Zeman PT, Musil J, Daniel R. High-temperature oxidation resistance of Ta-Si-N films with a high Si content. Surface and Coating Technology. 2006;**200**:4091-4096. DOI: 10.1016/j.surfcoat.2005.02.097

[51] Daniel R, Musil J, Zeman P, Mitterer C. Thermal stability of magnetron sputtered Zr-Si-N films. Surface and Coatings Technology. 2006;**201**(6):3368-3376. DOI: 10.1016/j.surfcoat.2006.07.206

[52] Dong Y, Zhao W, Li Y, Li G. Influence of silicon on the microstructure and mechanical properties of Zr-Si-N composite films. Applied Surface Science. 2006;**252**(14):5057-5062. DOI: 10.1016/j.apsusc.2005.07.050

[53] Pilloud D, Pierson JF, Pichon L. Influence of the silicon concentration on the optical and electrical properties of reactively sputtered Zr-Si-N nanocomposite coatings. Materials Science and Engineering B. 2006;**131**(1-3):36-39. DOI: 10.1016/j.mseb.2006.03.017

[54] Sandu CS, Medjani F, Sanjinés R, Karimi A, Lévy F. Structure, morphology and electrical properties of sputtered Zr-Si-N thin films: From solid solution to nanocomposite. Surface and Coatings Technology. 2006;**201**(7):4219-4223

[55] Pilloud D, Pierson JF, Takadoum J. Structure and tribological properties of reactively sputtered Zr-Si-N films. Thin Solid Films. 2006;**496**(2):445-449. DOI: 10.1016/j.tsf.2005.09.062

[56] Winkelmann A, Cairney JM, Hoffman MJ, Martin P, Bendavid A. Zr-Si-N films fabricated using hybrid cathodic arc and chemical vapour deposition: Structure vs. properties. Surface and Coatings Technology. 2006;**200**(14-15):4213-4219. DOI: 10.1016/j.surfcoat.2005.01.004

[57] Musil J, Daniel R, Zeman P, Takai O. Structure and properties of magnetron sputtered Zr-Si-N films with a high (≥25 at.%) Si content. Thin Solid Films. 2005;**478**(1-2):238-247. DOI: 10.1016/j.tsf.2004.11.190

[58] Martin PJ, Bendavid A, Cairney JM, Hoffman M. Nanocomposite Ti-Si-N, Zr-Si-N, Ti-Al-Si-N, Ti-Al-V-Si-N thin film coatings deposited by vacuum arc deposition. Surface and Coatings Technology. 2005;**200**(7):2228-2235. DOI: 10.1016/j.surfcoat.2004.06.012

[59] Pilloud D, Pierson JF, Marques AP, Cavaleiro A. Structural changes in Zr-Si-N films vs. their silicon content. Surface and Coatings Technology. 2004;**180-181**:352-356. DOI: 10.1016/j.surfcoat.2003.10.087

[60] Song ZX, Xu KW, Chen H. The characterization of Zr-Si-N diffusion barrier films with different sputtering bias voltage. Thin Solid Films. 2004;**468**(1-2):203-207. DOI: 10.1016/j.tsf.2004.04.037

[61] Song Z, Xu K, Chen H. The effect of nitrogen partial pressure on Zr-Si-N diffusion barrier. Microelectronic Engineering. 2004;**71**(1):28-33. DOI: 10.1016/j.mee.2003.08.006

[62] Zhou M, Nose M, Deguchi Y, Mae T, Nogi K. Influence of sputtering conditions on microstructure and mechanical properties of Zr-Si-N films prepared by radio-frequency-reactive sputtering. Journal of Vacuum Science & Technology A-Vacuum Surfaces and Films. 2003;**21**(5):1791-1795. DOI: 10.1116/1.1598976

[63] Nose M, Chiou WA, Zhou M, Mae T, Meshii M. Microstructure and mechanical properties of Zr-Si-N films prepared by rf-reactive sputtering. Journal of Vacuum Science & Technology A-Vacuum Surfaces and Films. 2002;**20**(3):823-828. DOI: 10.1116/1.1468657

[64] Veprek S, Zhang RF, Veprek-Heijman MGJ, Sheng SH, Argon AS. Superhard nanocomposites: Origin of hardness enhancement, properties and applications. Surface and Coatings Technology. 2010;**204**:1898-1906. DOI: 10.1016/j.surfcoat.2009.09.033

[65] Changjie F, En C, Mingsheng L, Namei W. Effects of Cu content on microstructure and high-temperature oxidation behavior of Ti-Al-Si-Cu-N nanocomposite films. Rare Metals Materials And Engineering. 2017;**46**(3):627-633

[66] Hsu CH, Huang WC, Lee YP, Ho WY. Effect of nitrogen atmosphere heat treatment on structure and wear behavior of CrAlSiN nanocomposite film. Surface and Coatings Technology. 2017;**320**:230-234. DOI: 10.1016/j.surfcoat.2017.01.040

[67] Adachi H, Wasa K. Thin films and nanomaterials. In: Wasa K, Kanno I, Kotera H, editors. Handbook of Sputtering Technology. 2nd ed. William Andrew; 2012. p. 3-39. DOI: 10.1016/B978-1-4377-3483-6.00001-2

[68] Rezaee S, Ghobadi N. Synthesis of Ag-Cu-Pd alloy thin films by DC-magnetron sputtering: Case study on microstructures and optical properties. Results in Physic.

2018;**9**:1148-1154. DOI: 10.1016/j.rinp.2018.04.029

[69] Novakovi M, Popovi M, Rakočevi Z, Bibi N. Structural, optical and electrical properties of reactively sputtered Cr_xN_y films: Nitrogen influence on the phase formation. Processing and Application of Ceramics. 2017;**11**(1):45-51. DOI: 10.2298/PAC1701045N

[70] Perez-Mariano J, Lau K-H, Sanjurjo A, Caro J, Casellas D, Colominas C. TiSiN nanocomposite coatings by chemical vapor deposition in a fluidized bed reactor at atmospheric pressure (AP/FBR-CVD). Surface and Coatings Technology. 2006;**201**(6):2217-2225. DOI: 10.1016/j.surfcoat.2006.03.029

[71] Velasco L, Olaya JJ, Rodil SE. Effect of Si addition on the structure and corrosion behavior of NbN thin films deposited by unbalanced magnetron sputtering. Applied Physics A: Materials Science & Processing. 2016;**122**(2):1-10. DOI: 10.1007/s00339-016-9639-0

[72] Cavaleiro A, Marques AP, Fernandes JV, Carvalho NJM, De Hosson JT. Evolution of the microstructure, residual stresses, and mechanical properties of W-Si-N coatings after thermal annealing. Journal of Materials Research. 2005;**20**(05):1356-1368. DOI: 10.1557/JMR.2005.0169

[73] Shah HN, Jayaganthan R, Pandey AC. Nanoindentation study of magnetron-sputtered CrN and CrSiN coatings. Materials and Design. 2011;**32**(5):2628-2634. DOI: 10.1016/j.matdes.2011.01.031

[74] Eriksson AO, Zhu J, Ghafoor N, Jensen J, Greczynski G, Johansson MP, et al. Ti-Si-C-N thin films grown by reactive arc evaporation from Ti 3SiC2 cathodes. Journal of Materials Research. 2011;**26**(7):874-881. DOI: 10.1557/jmr.2011.10

[75] Lee HY, Jung WS, Han JG, Seo SM, Kim JH, Bae YH. The synthesis of CrSiN film deposited using magnetron sputtering system. Surface and Coatings Technology. 2005;**200**:1026-1030. DOI: 10.1016/j.surfcoat.2005.02.006

[76] Vepřek S. The search for novel, superhard materials. Journal of Vacuum Science & Technology A-Vacuum Surfaces and Films. 1999;**17**(5):2401-2420. DOI: 10.1116/1.581977

[77] Seshan K. Handbook of Thin-Film Deposition Processes and Techniques: Principles, Methods, Equipment and Applications. 2nd ed. Noyes Publications; 2002

[78] Stenzel O. Optical Coatings. Verlag Berlin Heidelberg: Springer; 2014. 378 p. DOI: 10.1007/978-3-642-54063-9

[79] Alfonso JE, Olaya J, Cubillo G. Thin film growth through sputtering technique and its applications. In: Rubens M, Andreeta B, editors. Crystallization. Rijeka, Croatia: IntechOpen; 2012. pp. 397-432. DOI: 10.5772/35844

[80] Daniel R, Musil J. Novel Nanocomposite Coatings: Advances and Industrial Applications. CRC Press; 2013. p. 310

[81] Signore MA, Valerini D, Rizzo A, Tapfer L, Capodieci L, Cappello A. Investigation of the physical properties of ion assisted ZrN thin films deposited by RF magnetron sputtering. Journal of Physics D: Applied Physics. 2010;**43**:1-8. DOI: 10.1088/0022-3727/43/22/225401

[82] Vaz F, Carvalho P, Cunha L, Rebouta L, Moura C, Alves E, et al. Property change in ZrNxOy thin films: Effect of the oxygen fraction and bias voltage. Thin Solid Films. 2004;**469-470**:11-17. DOI: 10.1016/j.tsf.2004.06.191

[83] Kiahosseini SR, Afshar A, Larijani MM, Yousefpour M. Adhesion, microstrain, and corrosion behavior of ZrN-coated AZ91 alloy as a function of temperature. Journal of Materials Research. 2013;**28**(19):2709-2714. DOI: 10.1557/jmr.2013.241

[84] Soldán J, Musil J. Structure and mechanical properties of DC magnetron sputtered TiC/Cu films. Vacuum. 2006;**81**(4):531-538. DOI: 10.1016/j.vacuum.2006.07.013

[85] Veszelei M, Andersson K, Ribbing C-G, Järrendahl K, Arwin H. Optical constants and Drude analysis of sputtered zirconium nitride films. Applied Optics. 1994;**33**(10):1993

[86] Rizzo A, Signore MA, Mirenghi L, Serra E. Properties of ZrNx films with x>1 deposited by reactive radiofrequency magnetron sputtering. Thin Solid Films. 2006;**515**(4):1307-1313. DOI: 10.1016/j.tsf.2006.03.020

[87] Bhatt V, Chandra S. Silicon nitride films deposited by RF sputtering for microstructure fabrication in MEMS. Journal of Electronic Materials. 2009;**38**(9):1979-1989. DOI: 10.1007/s11664-009-0846-8

[88] Vaz F, Machado P, Rebouta L, Mendes J, Lanceros-Méndez S, Cunha L, et al. Physical and morphological characterization of reactively magnetron sputtered TiN films. Thin Solid Films. 2002;**420-421**:421-428

[89] Zhao X, Jin J, Cheng JC, Lee JW, Wu KH, Lin KC, et al. Structural and optical properties of zirconia thin films deposited by reactive high-power impulse magnetron sputtering. Thin Solid Films. 2014;**570**(PB):404-411. DOI: 10.1016/j.tsf.2014.05.060

[90] Ribeiro E, Rebouta L, Carvalho S, Vaz F, Fuentes GG, Rodriguez R, et al. Characterization of hard DC-sputtered Si-based TiN coatings: The effect of composition and ion bombardment. Surface and Coatings Technology. 2004;**188-189**(1-3 Spec. Iss):351-357. DOI: 10.1016/j.surfcoat.2004.08.020

[91] Colligon JS, Vishnyakov V, Valizadeh R, Donnelly SE, Kumashiro S. Study of nanocrystalline TiN/Si_3N_4 thin films deposited using a dual ion beam method. Thin Solid Films. 2005;**485**(1-2):148-154. DOI: 10.1016/j.tsf.2005.03.036

[92] Cui X, Jin G, Hao J, Li J, Guo T. The influences of Si content on biocompatibility and corrosion resistance of Zr-Si-N films. Surface and Coatings Technology. 2013;**228**:524-528. DOI: 10.1016/j.surfcoat.2012.04.060

[93] Liu H, Tang W, Hui D, Hei L, Lu F. Characterization of (Al, Si)N films deposited by balanced magnetron sputtering. Thin Solid Films. 2009;**517**(21):5988-5993. DOI: 10.1016/j.tsf.2009.03.173

[94] Musil J. Hard nanocomposite coatings: Thermal stability, oxidation resistance and toughness. Surface and Coatings Technology. 2012;**207**:50-65. DOI: 10.1016/j.surfcoat.2012.05.073

[95] Calderon V S, Cavaleiro A, Carvalho S. Chemical and structural characterization of ZrCNAg coatings: XPS, XRD and Raman spectroscopy. Applied Surface Science. 2015;**346**:240-247. DOI: 10.1016/j.apsusc.2015.03.161

[96] Brandes EA, Brook GB. Smithells Metals Reference. 7th ed. London: Butterworth-Heinemann; 2004

[97] Patscheider J, Zehnder T, Diserens M. Structure-performance relations in nanocomposite coatings. Surface and Coatings Technology. 2001;**146-147**:201-208

[98] Sanjinés R, Sandu CS. Chapter 19: Interfacial electron scattering in nanocomposite materials: Electrical measurements to reveal the

Nc-MeN/a-SiN_x nanostructure in order to tune macroscopic properties. New Trends and Developments. 2012: 483-503. DOI: 10.5772/51123

[99] Kertzman Z, Marchal J, Suarez M, Staia MH, Filip P, Kohli P, et al. Mechanical, tribological, and biocompatibility properties of ZrN-Ag nanocomposite films. Journal of Biomedical Materials Research Part A. 2008;**84**:1061-1067

[100] Tung HM, Huang JH, Tsai DG, Ai CF, Yu GP. Hardness and residual stress in nanocrystalline ZrN films: Effect of bias voltage and heat treatment. Materials Science and Engineering A. 2009;**500**(1-2):104-108. DOI: 10.1016/j.msea.2008.09.006

[101] Sandu CS, Harada S, Sanjinés R, Cavaleiro A. A unique approach to reveal the nanocomposite nc-MN/SiN-layer architecture of thin films via electrical measurements. Surface and Coating Technology. 2010;**204**:1907-1913. DOI: 10.1016/j.surfcoat.2009.11.045

[102] Ahmed MS, Munroe P, Jiang ZT, Zhao X, Rickard W, Z-f Z, et al. Corrosion behaviour of nanocomposite TiSiN coatings on steel substrates. Corrosion Science. 2011;**53**(11): 3678-3687. DOI: 10.1016/j.corsci.2011.07.011

[103] Ding JC, Wang QM, Liu ZR, Jeong S, Zhang TF, Kim KH. Influence of bias voltage on the microstructure, mechanical and corrosion properties of AlSiN films deposited by HiPIMS technique. Journal of Alloys and Compounds. 2019;**772**:112-121. DOI: 10.1016/j.jallcom.2018.09.063

[104] Ferreira CP, Castro M d MR d, Tentardini EK, Lins V de FC, Saliba PA. Silicon influence on corrosion resistance of magnetron sputtered ZrN and ZrSiN thin films. Surface Engineering. 2018;**0**(0):1-8. DOI: 10.1080/02670844.2018.1548100

Chapter 4

Silica-Reinforced Deproteinized Natural Rubber

Siti Salina Sarkawi

Abstract

Reinforcement effect of silica-silane system in deproteinized natural rubber (DPNR) is reported. The influence of mixing temperature and sulfur ranks in silane bis-triethoxysilylpropyl-polysulfide is compared between silica-DPNR and silica-natural rubber (NR). Dispersion morphology of silica and rubber-to-filler interaction in DPNR is visualized by atomic force microscopy and TEM network visualization, respectively. DPNR compound shows smaller influence of dump temperature and more constant mechanical properties as compared to NR compound. Establishment of the silica-silane-rubber coupling in DPNR tread compound results in an improvement in dynamic properties especially the lower tan δ at 60°C which indicates the lower rolling resistance of tire. DPNR shows better mechanical and dynamic properties as compared to NR.

Keywords: deproteinized natural rubber, silica, silane, tire, low rolling resistance

1. Introduction

The use of silica has become of growing importance in tires because of reduced fuel consumption in automotive transport and consequently preservation of the environment. The high-dispersion silica reinforcement was introduced in the early 1990, by Michelin in passenger tire tread rubbers, the so-called green tire [1], which offers approximately 30% lower rolling resistance, resulting in 5% fuel savings and lower carbon dioxide emission to the environment [2, 3].

The core of the high-dispersion silica technology is the nanoscale reaction of the 4–6 silanol groups per nm^2 on the surface of the 20–30 nm diameter primary silica particles with a coupling agent. This reaction reduces the hydrophilic character of the filler and increases its compatibility with the rubber polymer [2, 4, 5]. This reaction is due to take place in the same processing step as mixing of the tire compound and is very difficult to lead to completion [6]. The coupling agent eventually creates a chemical link between the primary silica particles and the rubber molecules during the later processing steps of vulcanization [7, 8].

Silica technology as it is used today employs solution-polymerized synthetic elastomers like solution styrene-butadiene rubber (sSBR) and solution butadiene rubber (BR). On the other hand, the great majority of rubber polymers used for carbon-black reinforced tire applications are emulsion polymers such as natural rubber and emulsion styrene-butadiene rubber (E-SBR). Research on reinforcement of the silica-silane system has been extensively investigated, and most of the early silica compound development involved natural rubber as the base polymer [9]. Up until now, natural rubber is not fully utilized in silica technology due to postulation of its ineffectiveness with silane coupling agent. An in-rubber study of the interaction of silica with proteins present in natural rubber shows that proteins compete with the coupling agents for reaction with the silica during mixing, making the silane less efficient for improving dispersion and filler-polymer coupling and consequently affecting the final properties of the compound [10].

Natural rubber is a linear, long-chain polymer with repeating isoprene units (C_5H_8) [11]. It has a density of 0.913 at 20°C and glass transition temperature of −72°C [12]. The average molecular weight (Mw) of commercial NR is $1–1.5 \times 10^6$, and number average molecular weights are $3–5 \times 10^5$ [13]. The commercial NR is produced from coagulation of latex which is tapped from *Hevea brasiliensis* or rubber tree. The composition of Hevea latex is mainly rubber hydrocarbon of 30–45% weight; nonrubber components for 3–5% and the balance is water. The nonrubber components are proteins, lipids, amino acids, amine, carbohydrates, inorganic materials, and minerals [14].

The fundamental structure of NR has been revealed by NMR studies as a linear rubber chain consisting of initiating terminal (ω-terminal), two trans-1,4 isoprene units, long sequence of 1000–3000 cis-1,4 isoprene units, and chain-end terminal (α-terminal) as shown in **Figure 1** [15–17]. The ω-terminal consists of mono- and diphosphate groups linked with phospholipids by hydrogen bond or ionic bond [18]. The α-terminal is a modified dimethylallyl unit linked with functional groups, which is associated with proteins to form cross-linking via hydrogen bonding. These functional groups at both terminals are presumed to play a role in the branching and gel formation in NR [19, 20]. These secondary structures play a significant role in the strain-induced crystallization of unvulcanized and vulcanized natural rubber [21].

Modification of natural rubber is carried out to improve the behavior of NR during rubber product processing and to improve the in-service performance of the products which can be outside its traditional applications. NR is modified through physical and chemical processes. The major modifications of NR are outlined in **Figure 2** [12]. The physical modification of NR incorporates additives and other compounding ingredients which are not chemically reacting with rubber. Examples of physical-modified form of NR are oil extended NR, thermoplastic NR, powdered NR, and deproteinized NR. The chemical modification of NR involves reacting the NR chains chemically by attachment of pendant functional groups, grafting of different polymers along the rubber chains or through intramolecular changes and bond rearrangement. The chemical-modified form of NR includes epoxidized NR (ENR) [22], hydrogenated NR, and poly-methyl methacrylate (PMMA)-grafted NR.

The first use of silica in truck tire treads based on NR had shown an improvement in tear properties in terms of cut and chip behavior, but the amount used was limited to 25 phr in order to avoid negative effects on tread wear [2]. Higher amounts of silica require coupling agents, which at that time were not used for

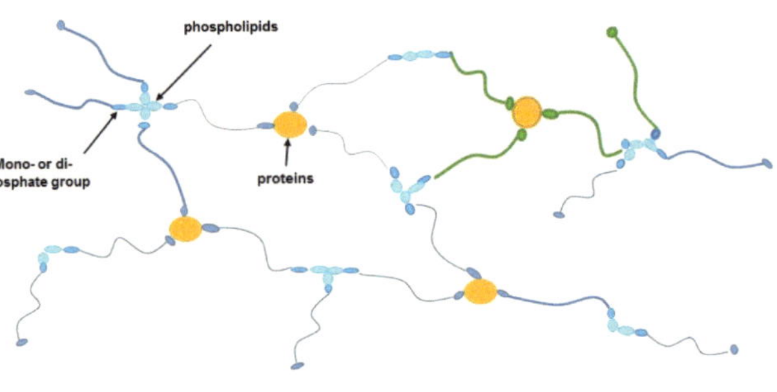

Figure 1.
A linear rubber chain structure with naturally occurring network associated with proteins and phospholipids.

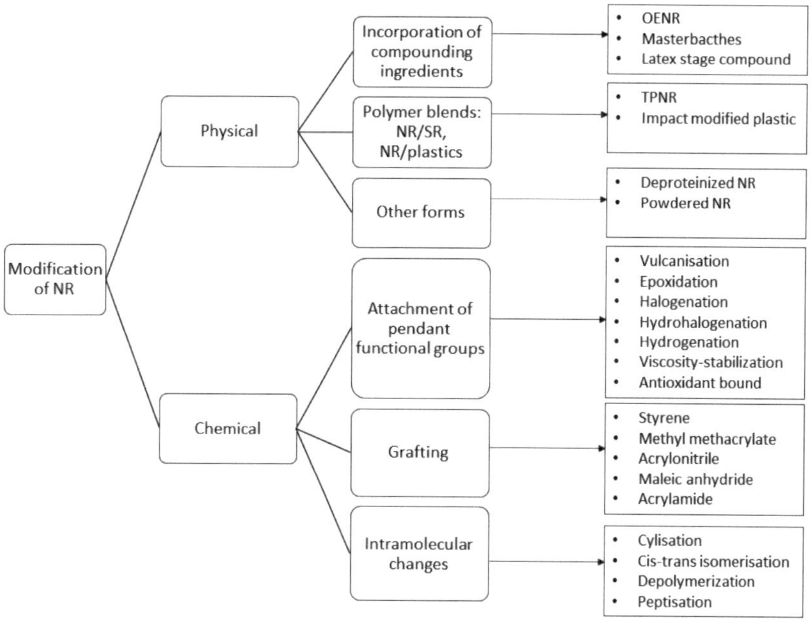

Figure 2.
Modification of natural rubber.

silica. From a study on optimization of the mixing conditions of silica-filled NR compounds with silane, an increase of mixing temperature and time for the silica modification with silane enhances the compatibility between silica and NR through a chemical bond between bis-triethoxysilylpropyl-tetrasulfide (TESPT) and the rubber [23]. The overall properties are dependent on the extent of this interaction [23]. The field tests with truck treads demonstrated that the rolling resistance can be improved by 30% when TESPT-modified silica is used in comparison to N220 carbon black. The tread wear index (abrasion resistance) decreases by no more than 5%, and wet traction shows little changes [24].

High reinforcement of properties for silica in NR and other rubbers needs to employ silane coupling agent [2, 7, 9, 25]. High reinforcing effect of silica is obtained with ENR without the use of coupling agents [26, 27] or reduced amount of coupling agent [28]. The vulcanizate properties of silica-filled ENR are comparable to carbon black compound at similar loading and are superior to a silica-filled NR without silane coupling agent (**Table 1**) [26, 29]. A comparison of physical properties of ENR-25—silica tread compound without coupling agent to sSBR/BR—silica compound is shown in **Table 2** [30]. In contrast with sSBR/BR, the ENR compound shows better reinforcement with silica without coupling agent. ENR exhibits both excellent wet grip and very low rolling resistance properties which is an attractive choice for tire tread compound for both passenger car and bus/truck applications [22, 31, 32].

Alternative approaches in incorporating silica into NR as nanocomposites are via in situ sol-gel technique, admicellar polymerization, and polymer-encapsulated silica [33]. Using sol-gel process, solid rubber is swollen in a silica precursor, e.g., tetraethyl orthosilicate (TEOS), followed by the sol-gel reaction, and silica content in the range of 15–22 wt % is achievable with stiffer and stronger tensile strength of 24 MPa. The in situ sol-gel technique on ENR with 3-aminopropyltriethoxysilane (APS) has been reported [34]. The ENR-APS sol-gel system with 28% sol-gel silica has higher tensile properties compared to a conventional ENR-sulfur-cured silica vulcanizate at 27% silica content with no silane coupling agent (**Table 3**) [35]. In

Properties	50 phr carbon black[a]			50 phr silica[b]		
	NR	ENR25	ENR50	NR	ENR25	ENR50
Tensile strength, MPa	29	26	25	24	21	27
Elongation at break, %	495	435	500	720	405	435
M300, MPa	12	12	13	6	13	13
Hardness, IRHD	65	69	73	69	67	68
Compression set 24 h/70°C, %	18	17	21	32	18	22
Akron abrasion (mm^3/500 rev)	21	14	11	63	15	14
Heat buildup, ΔT, °C	7	7	23	47	7	19

[a]*Rubber formulation in phr: rubber 100, filler, oil 5, ZnO 5, StA 2, antioxidant 2, S 2, MBS 1.5.*
[b]*Rubber formulation in phr: rubber 100, filler, oil 5, ZnO 5, StA 2, antioxidant 2, S 2, MBS 1.5, DPG 0.5.*

Table 1.
Comparison of physical properties of black and silica-filled ENR vulcanizates.

Properties	sSBR/BR*		sSBR/BR*	ENR
	Silica	Silica-TESPT	Black N234	Silica
Filler (phr)	80	80	67.5	50
Tensile strength, MPa	16	20	19	24
Elongation at break, %	780	515	600	550
M300, MPa	3.9	8.4	7.6	9.5
Hardness, IRHD	82	62	65	63
Angle tear, N/mm	39	36	43	40
Akron abrasion, vol loss, mm^3	41	13	12	<20
Dunlop resilience, %	44	55	38	67

**70/30.*

Table 2.
Comparison of physical properties of ENR-silica and sSBR/BR tread compounds.

Properties	ENR-sulfur vulcanizate (no silane)	ENR-APS sol-gel	ENR-APS sol-gel
Silica content (%)	27	28	27
Tensile strength, MPa	12	19	16
Elongation at break, %	390	290	300
M100, MPa	2	7	5

Table 3.
Comparison of tensile properties of ENR-silica sulfur cured with ENR-APS sol-gel system.

admicellar polymerization, a bilayer of surfactant (admicelle) is formed on the surface of silica and polymerization of the monomer in the admicelle resulting in an ultrathin layer of polymer covering the silica [33, 36, 37]. The silica modified with admicellar polymerization has superior performance compared to those reinforced with unmodified or silica modified with typical coupling agents [37].

2. Deproteinized natural rubber

Purification of natural rubbers from nonrubber components is possible especially the protein removal. The deproteinization process of natural rubber yields a rubber with reduced protein content significantly as well as low ash and volatile matter which is known as deproteinized natural rubber (DPNR). The deproteinization process can be carried out with natural rubber latex via enzymatic treatment [38–40] or urea treatment [41, 42]. The principle of deproteinization with enzymatic treatment is to hydrolyze the proteins in the natural rubber latex into water soluble which will be removed during washing process of manufacturing the rubber [41, 42].

Structural changes of NR branch points are proposed to occur with the deproteinization process [15]. This is based on the findings that a linear rubber chain contains two types of functional groups at both terminals. After deproteinization, the branch points formed by the functional groups associated with proteins at the α-terminal through hydrogen bonding decompose and leave the branch points from phospholipid at the α-terminal. The long-chain branching in the purified NR originated from the interaction of phospholipids which link the rubber chain together [43–45]. The phospholipids are associated together by the formation of a micelle structure.

The commercially available DPNR is Pureprena, which is produced by the Malaysian Rubber Board and licensed to Felda Rubber Industries Sdn Bhd. Pureprena is a purified form of natural rubber and has a very-low-nitrogen, ash, and volatile matter contents as well as being lighter in color (**Table 4**). DPNR is less prone to storage hardening than a normal NR grade. When compounded using an efficient vulcanization (EV) system, DPNR has low creep and stress relaxation, low water absorption, low compression set, and a more consistent modulus when subjected to variable humidity conditions [46]. DPNR gives superior rubber compounds with excellent dynamic properties which are suitable for engineering applications (**Table 5**) [46].

Properties	Specification of SMR 20	Specification of Pureprena
Dirt retained on 44 μm aperture (% wt)	0.20 max	0.01 max
Ash content, (% wt)	1.00 max	0.15 max
Nitrogen content, (% wt)	0.60 max	0.12 max
Volatile matter content, (% wt)	0.80 max	0.30 max

Table 4.
Comparison of raw rubber properties specification of DPNR (Pureprena) and natural rubber (SMR20).

Features	Applications
Low stress relaxation and low creep	Hydromounts, seals, joint rings, large shock absorbers, suspension bushes, helicopter rotor bearings
Low water absorption	Underwater applications, large shock absorbers
Good dynamic properties	Anti-vibration mountings, surge fenders
Low protein and low ash	Medical, pharmaceutical, and food applications

Table 5.
Areas of application of DPNR.

3. Carbon black reinforcement of DPNR

The reinforcing of DPNR by carbon black has been studied in the past [46, 47]. For mixing carbon black and natural rubber, a high reinforcement can be obtained with good dispersion of filler as there is strong physical interaction between the carbon black and the rubber. DPNR-CV is a grade of viscosity stabilized at between 60 and 70 Mooney units. The typical vulcanizate properties of DPNR-CV in comparable to SMR CV is shown in **Table 6** [46]. DPNR-CV gives lower water absorption and compression stress relaxation. **Table 7** shows the physical properties of DPNR filled with carbon black at 50 phr loading and cure using EV system [46]. Similar results are observed where DPNR gives lower water absorption and compression stress relaxation.

The recent evaluation of DPNR (Pureprena) shows that its physical properties are comparable to SMR CV and SMR 10 (**Table 8**). Comparison of DPNR with synthetic polyisoprene (IR) is reported where DPNR shows similar tensile properties and low water absorption to IR (**Table 9**) [48]. The rubber compound using DPNR with total nitrogen content adjusted in the range of 0.12–0.3% by weight exhibits an improved low hysteresis loss property, excellent abrasion resistance, and remarkably excellent tear resistance which is advantageous to be used for tire tread [49].

Properties	DPNR-CV	SMR CV	SMR L
Tensile strength, MPa	33	32	33
Elongation at break, %	570	550	520
M100, MPa	2.4	2.9	2.7
M300, MPa	11.5	13.3	14.5
Hardness, IRHD	60	63	59
Resilience, %	69	68	70
Volume swell in water: 7 day/23°C, %	0.6	0.9	0.9
Compression stress relaxation: 25% strain, 7 day/23°C, %	10.5	12.4	11.5

**Rubber formulation: rubber 100, N330 35, ZnO 5, StA 2, TBBS 0.7, S 2.25. Cheang KT, Rais AR, Basir KB. Deproteinised Natural Rubber (DPNR), Malaysian Rubber Board (MRB) Monograph No. 10, C&D Trading, Kuala Lumpur, (2003).*

Table 6.
Typical vulcanizate properties of DPNR-CV.

Properties	DPNR CV	SMR CV
Tensile strength, MPa	26	27
Elongation at break, %	520	480
M300, MPa	11.3	13.5
Resilience, %	63	67
Hardness, shore A	52	55
Volume swell in water: 7 day/23°C, %	0.9	1.3
Compression stress relaxation: %, 25% strain, 7 day/23°C	8.6	11.7

**Rubber formulation: rubber 100, N774 50, ZnO 5, ZEH 2, Permanex TQ 2, MOR 1.44, TBTD 0.6, S 0.6, cured at 150°C for 20 min. Cheang KT, Rais AR, Basir KB. Deproteinised Natural Rubber (DPNR), Malaysian Rubber Board (MRB) Monograph No. 10, C&D Trading, Kuala Lumpur, (2003).*

Table 7.
Physical properties of DPNR.

Properties	DPNR	SMR CV	SMR L	SMR10
Tensile strength, MPa	28	26	27	27
Elongation at break, %	590	560	570	570
M100, MPa	2.4	2.4	2.4	2.5
M300, MPa	12.0	12.3	12.4	12.3
Hardness, IRHD	71	68	69	70
Resilience, %	55	57	58	55
Abrasion resistance index (ARI), %	109	106	110	114

**Rubber formulation: rubber 100, N234 60, ZnO 3, StA 2, 6PPD 2, Wax 1, TDAE 8, CBS 1.5, S 1.5, cured at 150°C for 6 min.*

Table 8.
Physical properties of DPNR (Pureprena).

Properties	DPNR[#]	IR[##]	SMR CV
Tensile strength, MPa	24	23	26
Elongation at break, %	700	730	700
M100, MPa	0.8	0.8	0.8
M300, MPa	2.6	2.5	2.6
Hardness, IRHD	42	40	41
Resilience, %	84	84	85
Compression set, 22 h 70°C, %	12	5	11
Volume swell in water: 7 day/23°C, % 3 day/100°C, %	0.1 1.7	0.2 1.5	0.4 3.8

**Rubber formulation: rubber 100, N774 20, ZnO 4, ZEH 2, TMQ 2, accelerator 1.44, accelerator 2 0.7, S 0.7.*

[#]Pureprena.

[##]SK1-3.

Abd Rahim R, Zaeimoedin TZ, Muhamad AK, Sarkawi SS. New Process Deproteinised Natural Rubber (Pureprena): Raw Rubber, Processability and Basic Physical Properties, Malaysian J. Analytical Sciences. 2016; 20(5):1145-1152.

Table 9.
Comparison of physical properties of DPNR with IR.

4. Silica reinforcement of DPNR

The reinforcement of rubber by silica is different from the reinforcement by carbon black. The use of conventional silica has been limited to a white filler for colored rubber compounds [2] such as shoe soles, until the introduction of bifunctional silanes as coupling agents back in the 1970s [50]. In 1992, Michelin patented the silica-filled rubber compound for the production of all-season high-performance tires [1], which provide an excellent compromise of low rolling resistance, good traction on wet and snow-covered roads, as well as noise reduction. The rubber compound is a blend of high-vinyl solution SBR and high-cis BR filled with highly dispersible silica [51] at a loading of 80 phr and an organosilane coupling agent. Since then, a lot of work on reinforcement of silica in SBR-based passenger car tires has been investigated [2, 7, 24, 52–54].

The in situ treatment of silica with silane is commonly adopted for mixing silica with rubber rather than the use of pre-reacted silica-silane. Compared to mixing carbon black, the in situ mixing of silica-silane in rubber requires optimal silanization to be completed to obtain high reinforcement [7]. The silanization is influenced by the temperature as it is a chemical reaction. Hence, mixing temperature is the parameter of paramount importance in mixing silica and rubber in the presence of coupling agent such as TESPT [55, 56]. The temperature has a more dominant effect than time in the silica-TESPT reaction [23]. In order to achieve a sufficient degree of silanization, the temperature during mixing should be between 150 and 160°C. However, above 160°C either the coupling agent starts to prematurely react with the rubber matrix or the TESPT starts to donate sulfur; both result in pre-scorch of the compound. A mixing time of at least 10 minutes at 150°C is necessary to ensure complete coupling of the silica and the silane and that the reaction between the silica and the silane takes place primarily during the first mixing step [50]. In an investigation of cure characteristics of NR and TESPT in comparison with SBR, it clearly shows that NR starts to react with TESPT at a temperature of 120°C but SBR only reacts with TESPT at a higher temperature of 150°C [56].

4.1 Influence of mixing dump temperature in silica-DPNR compound

The properties of silica-filled compounds show a strong dependence on mixing temperature, since the silanization reaction and silica dispersion are key parameters in this system [25]. The optimal mixing conditions for silica-filled NR compounds with TESPT as a coupling agent are required to obtain high reinforcement in NR [56]. The influence of the dump temperature of the first mixing step on the vulcanization behavior of the DPNR-silica compounds as compared to the NR-silica compounds is illustrated in **Figure 3**. Dump temperature is the final temperature reached when the compound is discharged from the internal mixer. The compound formulation is described in **Table 10**. The vulcanization curves and behavior of the NR-silica compounds can be divided into three groups according to dump temperature. The first group encompasses the compounds with low dump temperatures (below 150°C), the middle group is represented by the compounds with moderate dump temperatures around 150–155°C, and the third group is the compounds with high dump temperatures (>155°C). At low dump temperature region, both NR-silica and DPNR-silica compounds display pronounced silica flocculation as evidenced by initial torque rise, high maximum torques, and long scorch times. At

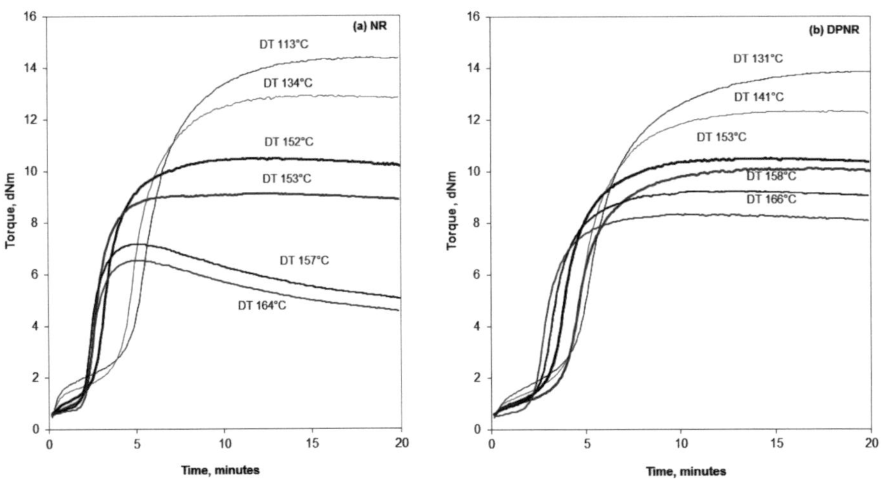

Figure 3.
Vulcanization behavior at 150°C of (a) the NR compounds and (b) the DPNR compounds, mixed till different dump temperatures (DT) in the first mixing stage.

Ingredients	Amount (phr)
Natural rubber (vary as stated in text)	100
Silica Ultrasil 7005	55
Silane TESPT	5
Zinc oxide	2.5
Stearic acid	1
TDAE oil	8
TMQ	2
Sulfur	1.4
CBS	1.7
DPG	2

Table 10.
Basic compounding formulation used for NR and DPNR.

moderate dump temperature region, no appearance of flocculation is observed for NR-silica and DPNR-silica compounds but at lower maximum torques and shorter scorch times compared to those of the compounds with low dump temperature. The maximum torque of the DPNR compound mixed at moderate temperature (around 150°C) is comparable with the NR compound at higher dump temperature region; both NR- and DPNR-silica compounds display a lower maximum torque and no sign of flocculation as the silanization reaction increases and results in a reduction of silica-silica interaction. Nonetheless, the decrease in maximum torque for the DPNR compound at higher dump temperature is much less compared to the NR-silica compound. The low-protein content in DPNR may attribute to better silanization in the compound which gives more filler-to-rubber linkages. Noticeably in the vulcanization curves for these high dump temperatures, there is clear indication of reversion for NR-silica compound. This reversion sign in vulcanization curve has also been reported in NR compound in the presence of TESPT at temperature higher than 150°C where the behavior is totally different as compared to SBR compound [56]. The DPNR-silica compounds show no sign of reversion at high dump temperature as seen in NR compounds. This suggests that a different network structure in the DPNR contributes to the thermal resistance of the cured compound.

A comparison of filler-filler interaction in DPNR-silica-TESPT compounds with NR-silica-TESPT as a function of mixing temperature is shown in **Figure 4**. A common way to measure the filler-filler interaction is calculating the Payne effect from the strain sweep of the dynamic testing or different in the storage modulus of the low or 0.56% and high strain or 100% while maintaining the frequency and temperature at constant value. Both NR- and DPNR-silica-TESPT show the decreasing trend of Payne effect with increasing dump temperature. This trend is due to the increase of silanization reaction with increasing temperature which results in the efficiency of the silane to hydrophobize the surface of the silica. High Payne effect of NR- and DPNR-silica compounds at low dump temperature is in agreement with the sign of flocculation observed in vulcanization curve. The results are also in agreement with other works reported on compound comprising TESPT showed a reduction of flocculation of silica when mixed at higher dump temperature [56, 57].

Figure 5 illustrates the effect of mixing temperature on Wolff's filler structure parameter, α_f. Wolff's filler structure parameter, α_f, is determined from the ratio

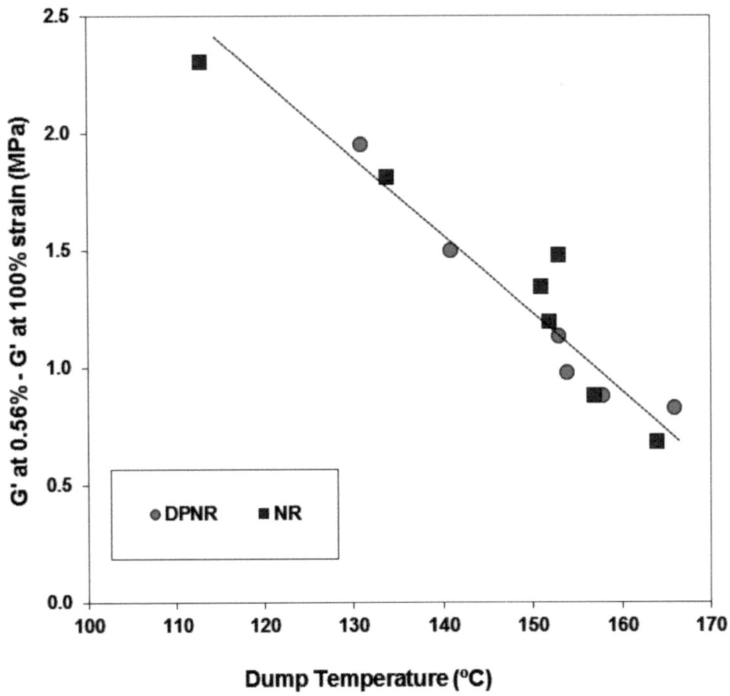

Figure 4.
Payne effect of silica compounds with silane TESPT as a function of dump temperature (□) NR and (•) DPNR.

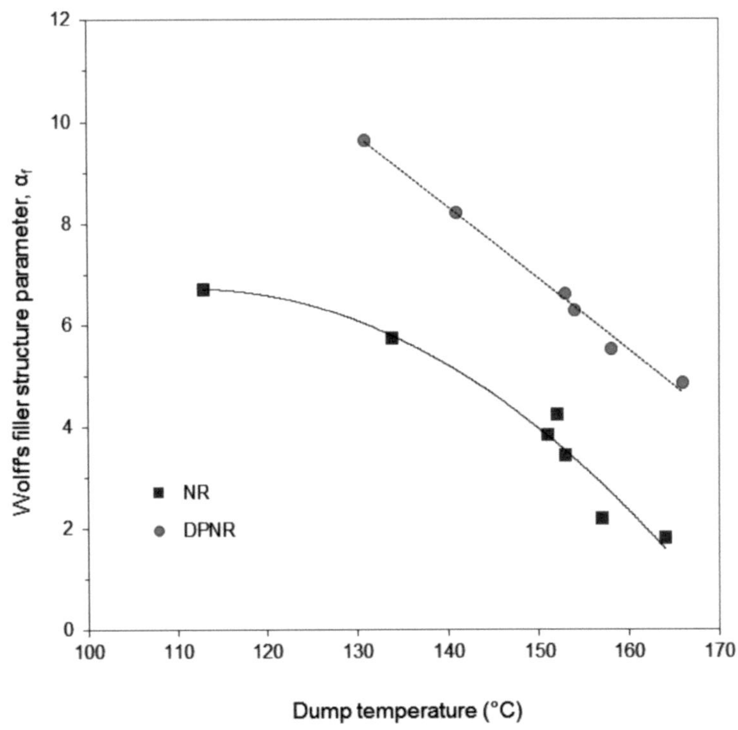

Figure 5.
Effect of dump temperature on Wolff's filler structure parameter, α_f of silica-reinforced compounds: (□) NR and (•) DPNR.

between the increase in rheometer torque during vulcanization of the filled compounds and that of the unfilled gum [24]:

$$\frac{D_{\max} - D_{\min}}{D^o_{\max} - D^o_{\min}} - 1 = \alpha_f \frac{m_f}{m_p} \tag{1}$$

where $D_{max} - D_{min}$ is the maximum change in torque for the filled rubber, $D^o{}_{max} - D^o{}_{min}$ is torque difference in unfilled rubber, and m_f/m_p is the ratio of filler to polymer by weight.

As observed earlier with the Payne effect, α_f is reduced with increasing dump temperature for the DPNR and NR compounds. Better hydrophobation leads to a decrease in silica-silica interaction and consequently results in reduced α_f. The DPNR compound shows a higher α_f than the NR compound, indicating a different type of filler and rubber network in the two compounds.

The filler-to-rubber interaction of silica-filled DPNR can be assessed from bound rubber content (BRC). Bound rubber is the layer of rubber polymer that covers the filler surface and unextracted by the solvent during swelling process. The bound rubber content is a combination of tightly bound rubber skin and loosely bound rubber shell [58]. These combination or referred to as total BRC can be measured using the bound rubber measurement in normal atmosphere. The tightly bound rubber skin can be measured from bound rubber measurement in an ammonia atmosphere which is referred to as the chemically BRC [59, 60].

The chemically and physically bound rubber content of silica-filled DPNR as compared to silica-filled NR is illustrated in **Figure 6**. The chemically bound rubber of both silica-filled NR and DPNR increases with increasing dump temperature up to 150°C. The increase of chemically bound rubber as a function of dump temperature for NR-silica-TESPT system has been reported [56, 25]. This can be explained by the higher rate of silanization with increasing mixing temperature. However, at 150°C, there is saturation in the amount of TESPT which has reacted and the surface of silica covered. Precipitated silica has about four to five silanol groups per nm^2. Hence, there is no increase in chemically bound rubber for compounds mixed till above 150°C dump temperature. Above 150°C, the chemically bound rubber slowly decreases for DPNR and NR. In comparison, DPNR has more chemically

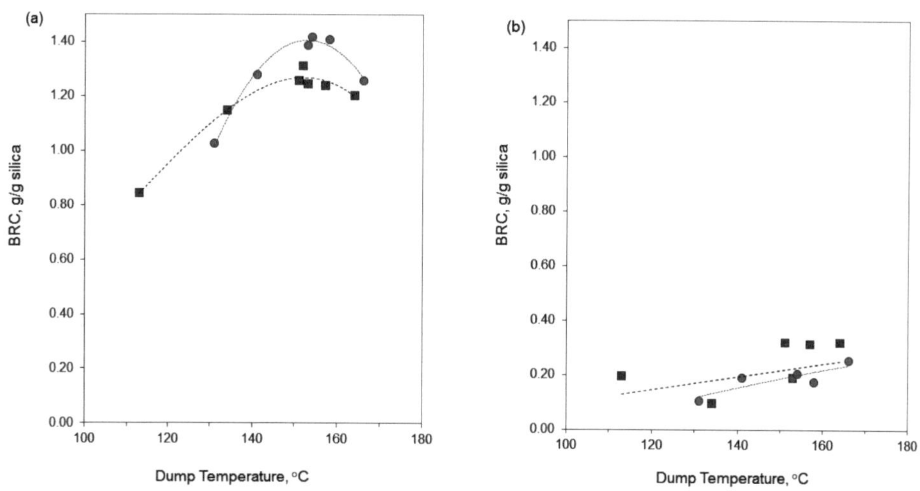

Figure 6.
Comparison of (a) chemically and (b) physically bound rubber content of silica compound containing TESPT at varying dump temperature: (□) NR and (•) DPNR.

bound rubber than NR, particularly at high dump temperature. In **Figure 6(b)**, the small increase in the physically bound rubber of the DPNR-silica and NR-silica compounds containing TESPT at higher dump temperature can be explained by the saturation of silica-TESPT coupling. Additional interactions above 150°C between the non-hydrophobized silica surfaces and rubber are physical of nature.

The effect of dump temperature on the physical properties of silica vulcanizates is depicted in **Figure 7**. At dump temperatures above 150°C, NR vulcanizate shows a clear reduction in tensile strength and is in good agreement with the occurrence of reversion and decrease in the maximum torque observed in the vulcanization curve. This shows that mixing temperature is of importance for the NR-silica compounds. This observation is also seen in the case of synthetic equivalent of NR, IR [61]. However, this reduction in tensile strength at high dump temperature is not seen for the low-protein DPNR-silica vulcanizates. The elongation at break for the NR and DPNR vulcanizates reduces slightly with increasing dump temperature.

A decrease in both moduli at 300% elongation (M300) and 100% elongation (M100) at higher dump temperatures is seen for the NR vulcanizates. In contrast, DPNR vulcanizates exhibit higher moduli with the increasing mixing dump temperature. The reduction in the tensile properties of NR at high mixing temperature is in good agreement with the occurrence of reversion and decrease in the maximum torque observed in their vulcanization curves. In addition, it has been reported that the reinforcement index (M300/M100) and tear resistance of NR-silica-TESPT vulcanizate is improved with increasing dump temperature to an optimum point

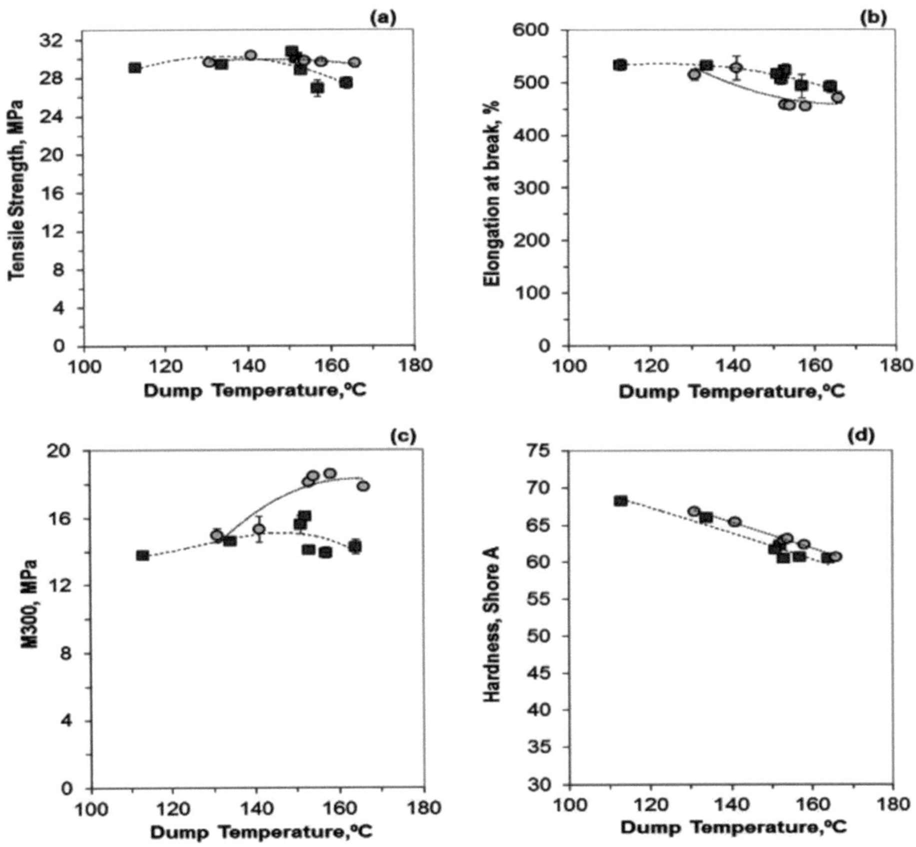

Figure 7.
Physical properties of silica vulcanizates: (□) NR and (•) DPNR; (a) tensile strength, (b) elongation at break, (c) modulus at 100% elongation (M100), (d) modulus at 300% elongation (M300).

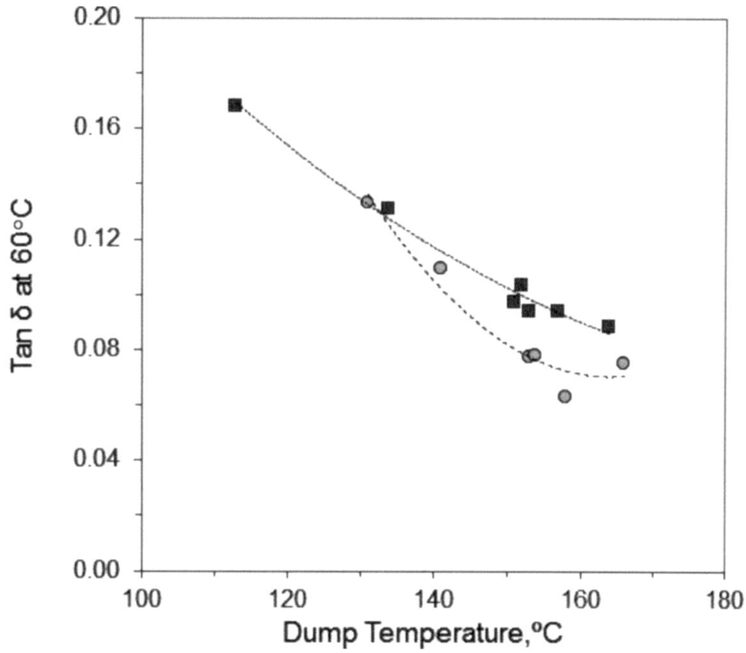

Figure 8.
The effect of dump temperature on tan delta at 60°C of silica compound: (□) NR and (•) DPNR.

of 135–150°C [56]. In contrast, reversion is not seen in the vulcanization curve of DPNR, and the effect of dump temperature on torque difference is also smaller. Another possible reason for the better properties achieved for DPNR is the difference in the naturally occurring networking as compared to NR. The network structure of purified DPNR has been proposed is associated with phospholipids linking both terminal end groups of the rubber chain via hydrogen bonding and ionic linkages, while the protein bonds are released because of deproteinization [43].

Commonly, the loss tan δ at 60°C of a cured compound is employed as indication for the rolling resistance of tires made thereof. The lower the tan δ at 60°C, the lower the rolling resistance expected in real tire performance. **Figure 8** illustrates indications of rolling resistance of the silica-filled vulcanizates. Both NR and DPNR vulcanizates show a strong decrease in tan δ at 60°C with increasing dump temperature. This must obviously be the result of more coupling of silica to the rubber with greater silanization efficiency at high temperatures. The DPNR vulcanizates exhibit the lowest tan δ at 60°C at high dump temperature. This actually relates well with the higher chemically bound rubber content of DPNR compared to the NR compound.

In general, DPNR shows better mechanical and dynamic properties as compared to NR. The decrease in the mechanical properties of NR vulcanizates mixed at high dump temperature correlates well with the reversion sign in vulcanization curve and decrease in the maximum torque. Interestingly, the DPNR-silica compounds show no sign of reversion at high dump temperature as seen in NR compounds and better retention of the properties.

4.2 Influence of silane coupling agents

Organofunctional silanes used for sulfur-cured rubber compounds can be categorized into the following groups [25]:

1. Di- and polysulfidic silanes: $[(RO_3)\text{—Si—}(CH_2)_3\text{—S}]_2\text{—}S_x$

2. Mercaptosilanes: $[(RO_3)—Si—(CH_2)_3—S]_2—SH$

3. Blocked mercaptosilanes: $[(RO_3)—Si—(CH_2)_3—S]_2—S—B$

where R = CH_3 or C_2H_5 or other groups, B = CN or $C_7H_{15}C{=}O$, and x = 0–8.

The comparison of reinforcing effect of different types of silane in DPNR-silica compound is given in **Table 11**. Silane Si-69 is an organosilane TESPT from Evonik. X50s is a blend of silane Si69 and N330 black in the ratio 1:1 by weight, and it is in solid form. All types of silanes give high reinforcement of silica in DPNR especially TESPT with excellent tensile properties and better abrasion resistance.

The commonly and effectively used silane coupling agent in rubber system is TESPT. The polysulfidic silanes like TESPT contribute additional sulfur to the compound unlike the other type of silanes. The influence of sulfur ranks in polysulfidic silane is compared between silica-DPNR and silica-NR. Bis-triethoxysilylpropyl-disulfide (TESPD) has an average sulfur rank of 2.2, while TESPT has an average sulfur rank of 3.83. The compound is prepared based on the earlier formulation (**Table 10**), and 4.4 phr of TESPD is used as the equivalent moles to 5 phr TESPT in

Properties	Polysulfidic silane		Mercaptosilane	Block mercaptosilane
Silane	Si-69	X50S	Si 363	NXT
phr	5	10	6.2	5
Tensile strength, MPa	31	31	28	30
Elongation at break, %	560	550	540	600
M100, MPa	2.8	2.9	2.9	2.3
M300, MPa	13.3	14.2	13.1	10.2
Hardness, IRHD	67	69	65	65
Resilience, %	67	68	74	64
Abrasion resistance index (ARI), %	120	123	121	108

**Rubber formulation: rubber 100, Zeosil 1165 55, ZnO 3, StA 2, TMQ 2, TDAE 8, DPG 1.1, CBS 1.5, S 1.5, cured at 150°C for 10 min.*

Table 11.
Physical properties of DPNR-silica with different type of silanes used.

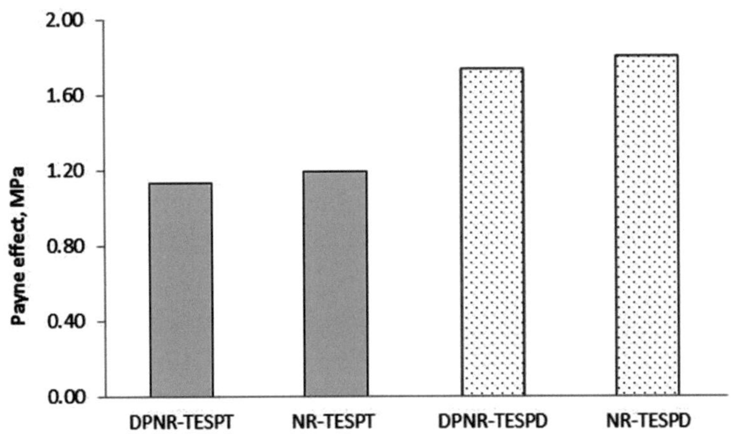

Figure 9.
Comparison of Payne effect between TESPT and TESPD compound.

ethoxy functionality. The TESPD and TESPT were compared on an equimolar basis with correction for the missing sulfur in the final mill mixing stage. The optimal loading of TESPT is at approximately 9.0% wt relative to the amount of silica [62]. In **Figure 9**, the Payne effect between NR and DPNR compounds is compared. DPNR compounds show slightly lower Payne effect compared to NR either with TESPT or TESPD. For both NR and DPNR compounds, the Payne effect of compound with TESPD is higher than those with TESPT. In this case, the reactivity of TESPT toward silica is higher in hydrophobizing silica surface than TESPD in both rubbers. The silanization silica with TESPD occurs at lower rate than TESPT. Similar observation has been reported where the TESPT-based NR compound shows lower filler-filler interaction and better silica dispersion than those of the TESPD compounds [63].

The comparison of BRC between NR- and DPNR-silica compounds both with and without silanes is shown in **Figure 10**. The use of silanes TESPT and TESPD in NR- and DPNR-silica compounds results in almost totally chemically BRC formation. The silanization of silica with TESPD or TESPT has successfully hydrophobize silica surface through silica-TESPT and silica-TESPD couplings. The silica-silane coupling is translated into reduction of the specific component of the surface energy of silica, γ^{s}_{sp}, and gives rise in the interaction of rubber and filler. DPNR-TESPT shows slightly higher chemically BRC than NR-TESPT-silica compound. In comparison, both DPNR and NR compounds with TESPT have higher BRC than compound with TESPD. This corresponds well with the lower Payne effect of the DPNR-silica compounds with TESPT as compared to with TESPD. For both DPNR- and NR-silica compounds without silane, only physically BRC, are formed as no chemically BRC is obtained after ammonia treatment. It is interesting to see that NR-silica without silane has higher physically BRC compared to DPNR-silica without silane. This relates back to the interaction of proteins with silica that provide filler-rubber

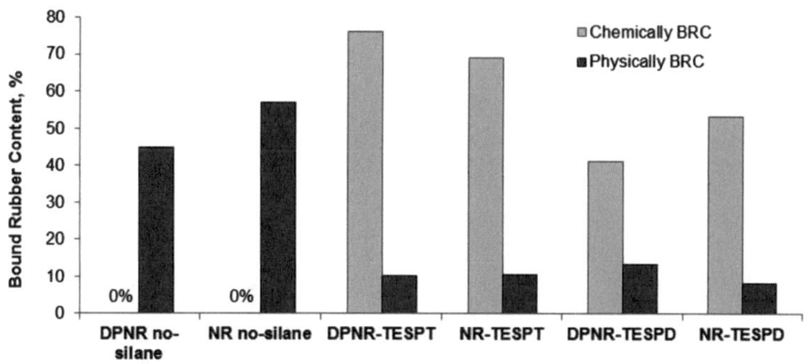

Figure 10.
Comparison of bound rubber content between TESPT and TESPD compounds.

Compounds	TS (MPa)	EB (%)	M100 (MPa)	M300 (MPa)	Hardness (shore A)	tan δ at 60°C
DPNR-silica-TESPT	29.7	460	3.5	18.1	63	0.077
NR-silica-TESPT	28.8	520	2.4	14.1	61	0.094
DPNR-silica-TESPD	29.4	520	3.8	14.9	70	0.139
NR-silica-TESPD	29.5	550	2.6	13.3	64	0.145

**Rubber formulation: rubber 100, Ultrasil7005 55, silane 5, ZnO 2.5, StA 1, TMQ 2, TDAE 8, DPG 2, CBS 1.7, S 1.4, cured at 150°C for 7 min.*

Table 12.
Comparison of physical properties between DPNR-TESPT and DPNR-TESPD vulcanizates.*

network in the NR-silica [64]. With reduced protein present in the DPNR, less network are formed in comparison to NR and this is reflected with lower BRC.

The comparison of physical properties between DPNR-silica-TESPT and DPNR-silica-TESPD vulcanizates is shown in **Table 12**. The tensile strength for all vulcanizates is not affected by the type of silane. The tensile properties of DPNR vulcanizate are comparable to NR vulcanizates. The indication of rolling resistance of the tread compounds can be observed from tan δ at 60°C from temperature sweep using dynamic mechanical testing. It is obvious that the use of TESPT gives lower tan δ at 60°C for DPNR-silica and NR-silica compounds as when compared to those using TESPD. A study on the role of different functionalities in silane on silica-filled NR has shown that the TESPT gives superior efficiency than TESPD due to the effect of sulfur donation by TESPT [65]. In addition, the DPNR vulcanizates generally give lower tan δ at 60°C than the NR vulcanizates which indicate an improvement in rolling resistance of tread compound made from DPNR.

4.3 Morphology of silica-reinforced DPNR

The dispersion morphology of silica-DPNR vulcanizates as compared to silica-NR vulcanizates by AFM morphology in the absence of silane coupling agent is illustrated in **Figure 11** [58]. The size of the silica aggregates in DPNR without silane is bigger than in the NR vulcanizate as seen from the height image at 1 × 1 μm. The phase image of DPNR-silica without silane shows smaller silica aggregates of 100nm size as dispersed in the matrix. The size of the silica aggregates in the DPNR vulcanizate is almost comparable to that in NR, although they seem to be closer together.

An improved micro-dispersion of silica in DPNR and NR vulcanizates with the use of TESPT is shown in AFM height images in **Figure 12** as compared to morphology without coupling agent. Primary particles of silica in the size of 50 nm are clearly visible in both DPNR and NR in addition to silica aggregates of approximately 100nm. The difference between DPNR and NR can be observed from the phase image. The distances between the silica aggregates of size 50–100 nm are

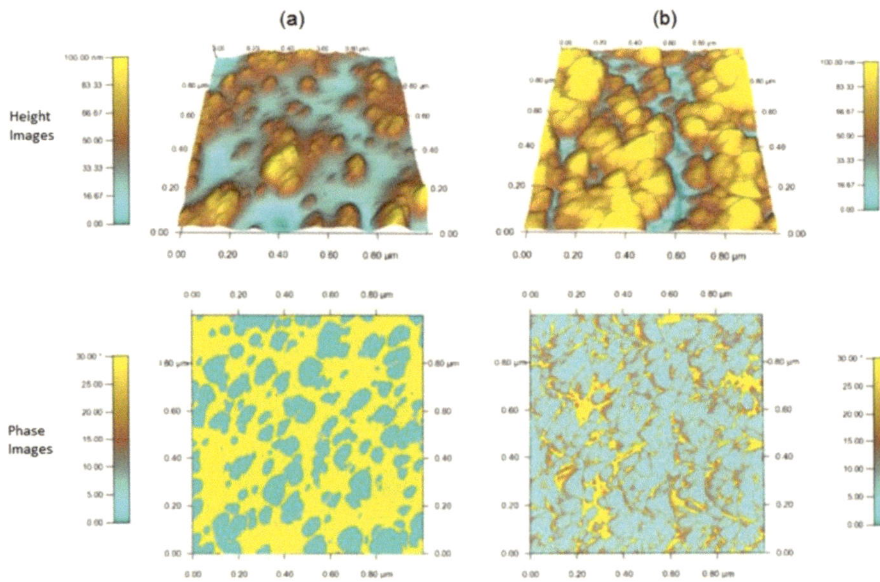

Figure 11.
Micro-dispersion of silica in NR and DPNR vulcanizates in the absence of silane coupling agent (a) NR-silica (b) DPNR-silica.

clearly visible in the NR-silica-TESPT vulcanizate. However, in DPNR-silica-TESPT, the distance between the aggregates looks a little smaller, indicating a somewhat better micro-dispersion. Besides, there is an intermediate region between the silica and rubber phases, suggesting the bound rubber layer [65–67].

Attempting to analyze the morphology of filler-to-rubber interactions in silica compounds at high loading, which in this study is 55 phr of silica, is difficult as the silica aggregates are very close together. In order to gain insight into the filler-to-rubber interaction, TEM network visualization was carried out where the vulcanizate was swollen in styrene, styrene polymerized, staining the rubber network and visualization using TEM [68].

TEM network visualizations of silica-filled NR and silica-filled DPNR vulcanizates without silane coupling agent are depicted in **Figure 13**. Silica aggregates of around 50–100 nm size can be seen as dark particles throughout the DPNR and NR vulcanizates. The rubber network can be visualized after the staining process, and the region is identified with mesh structure. Likewise, the region of

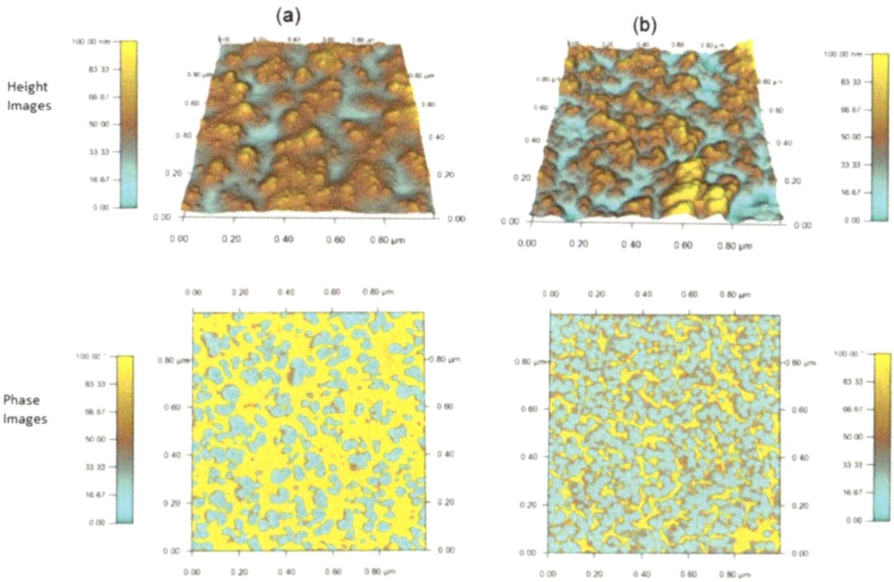

Figure 12.
Micro-dispersion of silica in NR and DPNR vulcanizates in the presence of silane coupling agent (a) NR-silica (b) DPNR-silica.

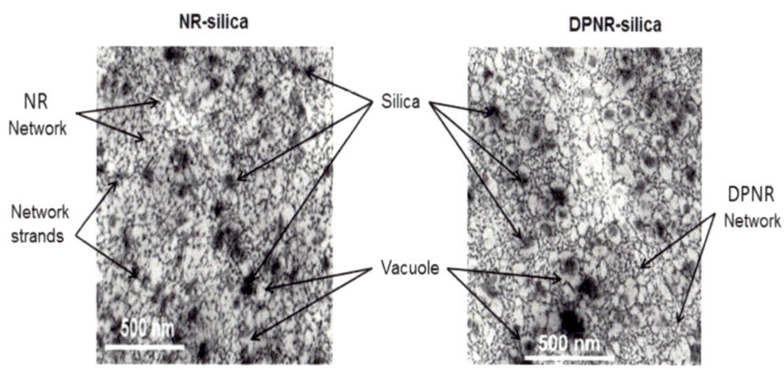

Figure 13.
Comparison of TEM network visualization micrographs of silica-filled NR and DPNR vulcanizates without coupling agent.

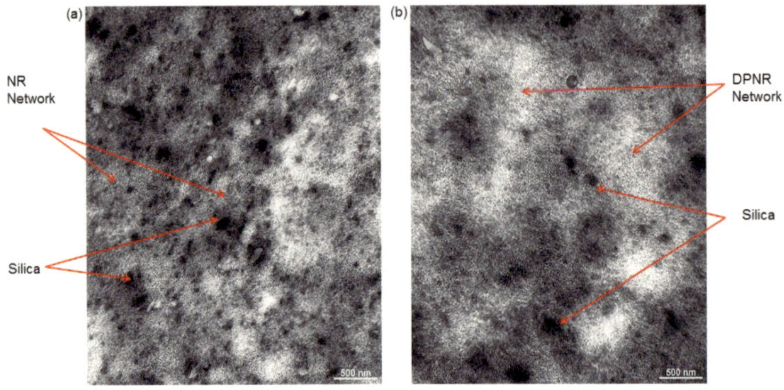

Figure 14.
Comparison of TEM network visualization micrographs of silica-filled (a) NR and (b) DPNR vulcanizates with silane coupling agent, TESPT.

polystyrene is the unstained part. Some silica aggregates in silica-filled NR are surrounded by voids or vacuoles, and some have connecting network strands to the NR network. This indicates there exist some bondings of silica to the rubber networks. This observation is totally different with the network visualization of the silica-filled DPNR. The silica aggregates in the DPNR system without silane have clear vacuoles surrounding them [62]. The vacuoles are formed through polymerization of styrene in the gap between the silica aggregates and the rubber chain. The formation of such vacuoles is due to a weak interface between silica particles and rubber chains [66]. The weak filler and rubber interaction in the system without silane is derived from the silica characteristic of surface energy with low dispersive component, γ_{sd}, which results in less adsorption of rubber chains to the surface of the silica. There are less vacuoles present in the NR vulcanizate as compared to the DPNR vulcanizate without silane, which suggests higher filler-to-rubber interactions in the former.

A comparison of the TEM network visualization between NR and DPNR vulcanizates with TESPT coupling agent included is shown in **Figure 14**. The presence of TESPT results in strong attachment of rubber chain to the surface of silica aggregates. No sign of vacuoles present in the system after network visualization. This is due to the establishment of chemical bonding between silica-TESPT-rubber in the compound during silanization and vulcanization. In addition, the size of silica aggregates in both NR and DPNR with TESPT is smaller than those without TESPT. This relates to the results of low Payne effect and high chemically BRC for NR and DPNR reinforced with silica and silane system. In addition, the vulcanizates with silanes exhibit denser rubber network compared to those without silane. This is in agreement with the results of cross-link density where the silica-filled NR and DPNR with TESPT have higher cross-link density compared to those without silane due to sulfur released from TESPT [62].

5. Conclusions

Deproteinized natural rubber is a purified form of natural rubber, which is characterized by its very low nitrogen, ash, and volatile matter contents compared to the commercial natural rubber. Similar to natural rubber, DPNR can be reinforced with carbon black as well as silica according to the intended applications.

DPNR reinforced with carbon black filler has low creep and stress relaxation, low water absorption, low compression set, and a more consistent modulus. Carbon black-filled DPNR exhibits excellent dynamic properties which are suitable for engineering applications.

Reinforcement of DPNR by silica filler is different from carbon black as silane coupling agent is required. Mixing temperature has a dominant effect in mixing DPNR and silica as well as NR and silica in the presence of a coupling agent. NR suffers some loss of dynamic and mechanical properties due to thermo-mechanical degradation when mixed at above optimum mixing temperature of approximately 155°C. On the other hand, DPNR shows a smaller influence of dump temperature and more constant mechanical properties. Higher silica-silane-rubber coupling in DPNR-silica-TESPT compounds is shown with improvement in the dynamic properties especially the lower tan δ at 60°C which indicates the lower rolling resistance of tire. The dispersion morphology of silica and rubber-to-filler interaction in DPNR is elucidated by atomic force microscopy and TEM network visualization.

Author details

Siti Salina Sarkawi
Malaysian Rubber Board, Rubber Technology Centre, RRIM Research Station, Sg Buloh, Selangor, Malaysia

*Address all correspondence to: ssalina@lgm.gov.my

References

[1] Rauline R. FR Patent 9102433. Compagnie Generale des Etablissements: Michelin-Michelin & Cie; 1992

[2] Meon W, Blume A, Luginsland HD, Uhrlandt S. Silica and silane. In: Rodgers B, editor. Rubber Compounding Chemistry and Applications. New York: Marcel Dekker Inc; 2004

[3] Neubauer J. Improving Key Tire Performance, Tire Technology International. 2009. pp. 10-12

[4] Gorl U, Hunsche A, Muller A, Koban HG. Investigations into the silica/silane reaction system. Rubber Chemistry and Technology. 1997;**70**:608-623

[5] Blume A. Analytical Properties of Silica—A Key for Understanding Silica Reinforcement; Spring, Chicago, USA: ACS Rubber Division Meeting, Paper No. 73; 1999

[6] Luginsland HD, Frohlich J, Wehmeier A. Influence of different silanes on the reinforcement of silica-filled rubber compounds. Rubber Chemistry and Technology. 2002;**75**:563-579

[7] Noordermeer JWM, Dierkes WK. Silica-filled rubber compounds. In: White J, De SK, Naskar K, editors. Rubber Technologist's Handbook. Vol. 2. Shawbury, Shrewbury, Shropshire Smithers: Rapra Technology; 2008. pp. 59-95

[8] Ten Brinke JW, Van Swaaij PJ, Reuvekamp LAEM, Noordermeer JWM. The influence of silane sulphur rank on processing of a silica-reinforced tire tread compound. Kautschuk Gummi Kunststoffe. 2002;**55**:244-254

[9] Sarkawi SS, Dierkes WK, Noordermeer JWM. A Review on Silica Reinforcement in Natural Rubber. Monograph No. 22; Malaysian Rubber Board, 2011.

[10] Sarkawi SS, Dierkes WK, Noordermeer JWM. The influence of non-rubber constituents on performance of silica reinforced natural rubber compounds. European Polymer Journal. 2013;**49**:3199-3209

[11] Barlow FW. Rubber Compounding: Principles, Materials and Techniques. New York: Marcel Dekker; 1988. pp. 9-28

[12] Mathew NM. Natural rubber. In: De SK, White JR, editors. Rubber Technologist's Handbook. Shawbury, Shrewbury, Shropshire: Rapra Technology Limited; 2001. pp. 11-46

[13] Barlow FW. General purpose elastomers—A natural rubber. In: Baranwal KC, Stephens HL, editors. Basic Elastomer Technology. 1st ed. Akron: The Rubber Division American Chemical Society; 2001. pp. 235-258

[14] Othman A, Hepburn C. Influence of non-rubber constituents on elastic properties of natural rubber vulcanizates. Plastics, Rubber, and Composites Processing and Applications. 1993;**19**:185

[15] Tarachiwin L, Sakdapipanich J, Tanaka Y. Structure and origin of long-chain branching and gel in natural rubber. Kautschuk Gummi Kunststoffe. 2005;**58**:115-122

[16] Eng AH, Kawahara S, Tanaka Y. Trans-isoprene units in natural rubber. Rubber Chemistry and Technology. 1993;**67**:159-168

[17] Amnuaypornsri S, Sakdapipanich J, Toki S, Hsiao BS, Ichikawa N, Tanaka Y. Strain-induced crystallization of

natural rubber: Effect of proteins and phospholipids. Rubber Chemistry and Technology. 2008;**81**:753-766

[18] Sakdapipanich J. Structural characterization of natural rubber based on recent evidence from selective enzymatic treatments. Journal of Bioscience and Bioengineering. 2003;**103**:287-292

[19] Sakdapipanich KNJ, Tanaka DMY. Structure of branch-points in natural rubber. Kautschuk Gummi Kunststoffe. 2008;**61**:518-522

[20] Tarachiwin L, Sakdapipanich J, Ute K, Kitayama T, Bamba T, Fukusaki E, et al. Structural characterization of α-terminal group of natural rubber. 1. Decomposition of branch-points by lipase and phosphatase treatments. Biomacromolecules. 2005;**6**:1851-1857

[21] Amnuaypornsri S, Nimpaiboon A, Sakdapipanich J. Role of phospholipids and proteins on gel formation and physical properties of NR during accelerated storage. Kautschuk Gummi Kunststoffe. 2009;**62**:88-90

[22] Gelling IR, Porter M. Chemical modification of natural rubber. In: Roberts AD, editor. Natural Rubber Science and Technology. Oxford: Oxford University Press; 1988. pp. 359-456

[23] Wolff S. Optimization of silane-silica OTR compounds—1. Variation of mixing temperature and time during the modification of silica with bis-(3-triethoxysilylpropyl)-tetrasulfide. Rubber Chemistry and Technology. 1982;**55**:967-789

[24] Wolff S. Chemical aspects of rubber reinforcement by fillers. Rubber Chemistry and Technology. 1996;**69**:325-346

[25] Sarkawi SS, Kaewsakul W, Sahakaro S, Dierkes WK, Noordermeer JWM. A review on reinforcement of natural rubber by silica fillers for use in low rolling resistance tires. Journal of Rubber Research. 2015;**18**(4):203-233

[26] Gelling IR. Epoxidised natural rubber. Journal of Natural Rubber Research. 1991;**6**(3):184-205

[27] Baker CSL, Gelling IR, Newell R. Epoxidized natural rubber. Rubber Chemistry and Technology. 1984;**58**:67-85

[28] Kaewsakul W, Sahakaro K, Dierkes WK, Noordermeer JWM. Cooperative effects of epoxide functional groups on natural rubber and silane coupling agents on reinforcing efficiency of silica. Rubber Chemistry and Technology. 2014;**87**:291-310

[29] Epoxidised Natural Rubber (ENR); Rubber Research Institute of Malaysia. Revised Ed. Kuala Lumpur: Raziq Prinkraf Sdn Bhd; 1984

[30] Chapman AV. Natural rubber and NR-based polymer: Renewable materials with unique properties. In: Proceeding of 24th International HF Mark Symposium; 15-16 November 2007; Vienna. 2007

[31] Martin PJ, Brown P, Chapman AV, Cook S. Silica-reinforced epoxidized natural rubber tire treads-performance and durability. Rubber Chemistry and Technology. 2015;**88**(3):390-411

[32] Martin PJ, Chapman AV, Cook S, Davies RT, Kepas-Suwara A. Green tire tread compounds with improved dispersion based on silica-reinforced epoxidized natural rubber. In: Proceeding of Fall 180th Technical Meeting of the Rubber Division of the American Chemical Society; 11-13 October 2011; Cleveland. 2011. Paper #36. ISSN 1547-1977

[33] Hashim AS, Ong SK. Silica reinforcement in natural rubber in

micro and nano length scales. In: Thomas S, Maria HJ, Joy J, Chan CH, Pothen LA, editors. Natural Rubber Materials. Vol. 2: Composites and Nanocomposites. Cambridge, UK: Royal Society of Chemistry; 2013. pp. 205-219

[34] Hashim AS, Kohjiya S, Ikeda Y. Moisture cure and in-situ silica reinforcement of epoxidized natural rubber. Polymer International. 1995;**38**:111-117

[35] Hashim AS, Kawabata N, Kohjiya S. Silica reinforcement of epoxidized natural rubber by the sol-gel method. Journal of Sol-Gel Science and Technology. 1995;**5**:211-218

[36] Nontasorn P, Chavadej S, Rangsunvigit P, O'Havior JH, Chaisirimahamorakot S, Na-Ranong N. Admicellar polymerization modified silica via a continuous stirred-tank reactor system: Comparative properties of rubber compounding. Chemical Engineering Journal. 2005;**108**:213-218

[37] Poochai C, Pae-on P, Pongpayoon T. Polyisoprene-coated silica/natural rubber composite. World Academy of Science, Engineering and Technology. 2010;**65**:969-973

[38] Chin PS, Chang WP, Lau CM, Pong KS. Deproteinised natural rubber (DPNR). In: Proceeding of RRIM Planters Conference. 1974

[39] Chang WP, Lau CM, Nambiar J. Deproteinised natural rubber from field latex. In: Proceedings of RRIM Planters Conference. 1977

[40] Cheang KT, Fong CS, Lian LC. A new and improved process of deproteinised natural rubber production. In: Proceedings of RRIM Growers Conference. 1987

[41] Kawahara S, Klinkai W, Kuroda H, Isono Y. Removal of proteins from natural rubber with urea. Polymers for Advanced Technologies. 2004;**15**:181-184

[42] Klinklai W, Saito T, Kawahara S, Tashiro K, Suzuki Y, Sakdapipanich J, et al. Hyperdeproteinized natural rubber prepared with urea. Journal of Applied Polymer Science. 2004;**93**:555-559

[43] Amnuaypornsri S, Tarachiwin L, Sakdapipanich J. Character of long-chain branching in highly purified natural rubber. Journal of Applied Polymer Science. 2009;**115**:3645-3650

[44] Rojruthai P, Tarachiwin L, Sakdapipanich JT, Toyonaka KU, Tanaka Y. Phospholipids associated with natural rubber molecules inducing long-chain branching, Part I. Kautschuk Gummi Kunststoffe. 2009;**62**:227-232

[45] Rojruthai P, Tarachiwin L, Sakdapipanich JT, Tanaka Y. Phospholipids associated with natural rubber molecules inducing long-chain branching, Part II. Kautschuk Gummi Kunststoffe. 2009;**62**:399-404

[46] Cheang KT, Rais AR, Basir KB. Deproteinised natural rubber (DPNR). In: Malaysian Rubber Board (MRB) Monograph No. 10, C and D Trading; Kuala Lumpur. 2003

[47] Kadir AA. Advances in natural rubber production. Rubber Chemistry and Technology. 1994;**67**:537-548

[48] Abd Rahim R, Zaeimoedin TZ, Muhamad AK, Sarkawi SS. New process deproteinised natural rubber (pureprena): Raw rubber, processability and basic physical properties. Malaysian Journal of Analytical Sciences. 2016;**20**(5):1145-1152

[49] Kondou H. Natural rubber, rubber composition and pneumatic tire. Bridgestone Corporation, US 7427646 B2; 2008

[50] Thurn F, Wolff S. Neue organosilane fur die reifenidustrie. Kautschuk Gummi Kunststoffe. 1975;**28**:733-739

[51] Chevallier Y, Morawski JC. EU Patent 0157701, Rhone-Poulenc Chimie; 1984

[52] Bomal Y, Touzet S, Barruel R, Cochet P, Dejean B. Developments in silica usage for decreased tire rolling resistance. Kautschuk Gummi Kunststoffe. 1997;**50**:434-439

[53] Luginsland HD. A Review on the Chemistry and the Reinforcement of the Silica-Silane Filler System for Rubber Applications. Aachen, Germany: Shaker Verlag; 2002

[54] Daudey S, Guy L. High performance silicas reinforced elastomers: From standard technology to advanced solutions. In: Proceedings of the Fall 180th ACS Rubber Division Meeting, Cleveland, Ohio, USA. 2011. Paper No. 37

[55] Reuvekamp LAEM, ten Brinke JW, van Swaaij PJ, Noordermeer JWM. Effects of time and temperature on the reaction of TESPT silane coupling agent during mixing with silica filler and tire rubber. Rubber Chemistry and Technology. 2002;**75**:187-198

[56] Kaewsakul W, Sahakaro S, Dierkes WK, Noordermeer JWM. Optimization of mixing conditions for silica-reinforced natural rubber tire tread compounds. Rubber Chemistry and Technology. 2012;**85**:277-294

[57] Lin CJ, Hergenrother WL, Alexanian E, Bohm GGA. On the filler flocculation in silica-filled rubbers. Part 1. Quantifying and tracking the filler flocculation and polymer-filler interactions in the unvulcanised rubber compounds. Rubber Chemistry and Technology. 2010;**75**:865-890

[58] Sarkawi SS, Dierkes WK, Noordermeer JWM. Morphology of silica-reinforced natural rubber: The effect of silane coupling agent. Rubber Chemistry and Technology. 2015;**88**:359-372

[59] Polmanteer KE, Lentz CW. Reinforcement studies-effect of silica structure on properties and crosslink density. Rubber Chemistry and Technology. 1975;**48**:795-809

[60] Wolff S, Wang MJ, Tan EH, Part VII. Study on bound rubber. Rubber Chemistry and Technology. 1992;**66**:163-177

[61] Sarkawi SS, Dierkes WK, Noordermeer JWM. Reinforcement of natural rubber by precipitated silica: The influence of processing temperature. Rubber Chemistry and Technology. 2014;**87**:103-119

[62] Kaewsakul W, Sahakaro K, Dierkes WK, Noordermeer JWM. Optimization of rubber formulation for silica reinforced natural rubber compound. Rubber Chemistry and Technology. 2013;**86**:313-329

[63] Kaewsakul W. Silica-reinforced natural rubber for low rolling resistance, energy-saving tires [thesis], Chapter 6. Zutphen: Wohrmann Print Service; 2013

[64] Sarkawi SS, Dierkes WK, Noordermeer JWM. Elucidation of filler-to-filler and filler-to-rubber interactions in silica reinforced natural rubber. European Polymer Journal. 2013;**49**:3199-3209

[65] Nakajima K, Nishi T. In: Bhowmick AK, editor. Chapter 21: Current Topics in Elastomers Research. New York: CRC Press; 2008

[66] Bielinski D, Slusarski L, Dobrowolski O, Dryzek E. Studies of filler agglomeration by atomic force microscopy and positron annihilation spectroscopy. Part 1: Silica filled

compounds. Kautschuk Gummi Kunststoffe. 2004;**57**:579-586

[67] Bielinski D, Slusarski L, Dobrowolski O, Dryzek E. Studies of filler agglomeration by AFM and PAS. Part II: Carbon black mixes. Kautschuk Gummi Kunststoffe. 2005;**58**:239-245

[68] Ladouce-Stelandr L, Bomal Y, Flandin L, Labarre D. Dynamic mechanical properties of precipitated silica filled rubber: Influence of morphology and coupling agent. Rubber Chemistry and Technology. 2003;**76**:145

Chapter 5

New Material for Si-Based Light Source Application for CMOS Technology

Luong Thi Kim Phuong

Abstract

In this chapter, an approach to enhance the radiative recombination of the Ge film grown on the Si substrate is presented. The Ge band gap structure could be modified by applying a tensile strain and high n-doping in the Ge epilayers. It thus becomes a direct band gap material with high photoluminescence efficiency which is compatible with mainstream silicon technology. The interdiffusion effect between Ge film and Si substrate is also mentioned in this section. We proposed a new method to suppress the Si/Ge interdiffusion to reduce the effect of Si atoms on the optical property of Ge film due to Si presence.

Keywords: Ge, Si technology, tensile strain, n-doping, photoluminescence, interdiffusion

1. Introduction

During the last decades, demands for increased performance of electronic devices have led to an extraordinary course of miniaturization of Complementary Metal Oxide Semiconductor (CMOS) devices. As this scaling law may soon reach its limit, researchers in modern microelectronics try to find solutions to extend Moore's law by using more conventional principles. Two major areas of research have been envisaged. The first aims to find solutions that will rely on the exiting existing technology and physical principles but on a smaller scale. As such, bottom-up nanoscale structures, such as semiconducting nanowires, carbon nanotubes or graphene, have been proposed as potential candidates for complementing Si technology. The second explores new concepts, which rely on introduction of novel functions, such as the development of on-chip optical interconnects or the introduction of spintronic devices in which not only the electron charge but also the electron spin that can carry information. On-chip optical interconnects imply integration of all the optical devices with the silicon microelectronic devices on silicon chips and they require an efficient Si-based optical source. In addition to the development of new generations of high-performance computation and communication systems, an integrated efficient optical source on silicon allows to bridge the gap between the microelectronic and optoelectronic industries. A Si-based optical source would allow developing silicon photonics and its diffusion toward extended new markets. It would also allow opening new possibilities with a strong economic potential by developing cost-effective devices intended for a large public diffusion. To date, numerous approaches have been proposed to realize silicon-integrated optical sources, such as porous Si, epitaxial semiconducting silicides, Er-doped Si, Si nanocrystals or Ge/Si self-assembled quantum dots. However, all above approaches are challenged by the lack of enough gain to surpass materials losses to achieve net gain for laser action.

Semiconductor diode lasers are conventionally based on direct band gap materials due to the efficient radiative recombination of direct gap transitions. As discrete devices, direct gap III-V semiconductor lasers have achieved great success in many important applications, from telecommunications to DVD players. On the other hand, indirect semiconductors such as Si and Ge are traditionally considered unsuitable for laser diodes because the radiative recombination through indirect transition is inefficient as a result of a phonon-assisted process. However, compared to Si, pure Ge displays unique optical properties, the direct (Γ) valley of its conduction band is only 0.14 eV above the indirect (L) valleys at room temperature while it is larger than 2 eV in Si. In addition, while Si cannot be used for photodetector due to its transparency at near infrared wavelength band, Ge has strong direct band gap absorption below 1.55 μm and high-speed Ge photodetectors have been demonstrated. Of particular interest, it has been shown that application of a tensile strain in Ge allows lowering the energy difference between the Γ zone center valley and the indirect L valley. The tensile strain also lifts the degeneracy between heavy hole and light hole valence bands. On the other hand, n-type doping of Ge leads to a more efficient population of the zone center Γ valley and thus enhances optical recombination at the Brillouin zone center. The combination of both effects is expected to lead to significant optical gains at room temperature and to the demonstration of a laser.

Tensile strain can be induced in Ge via several approaches: applying external mechanical stress, growing Ge on a larger lattice parameter substrate, such as InGaAs or GeSn buffer layers, or use of thermal mismatch between Ge and Si. In the frame of this chapter, we have chosen to study tensile-strained and n-doped Ge layers epitaxially grown on Si. The motivation of our study is that this system is compatible with the mainstream Si technology. Concerning to the n-doping process, while in CVD, PH_3 molecules are currently used as a precursor for n-doping, by means of MBE we could implement a specific doping cell using a GaP decomposition source to produce diphosphorus (P_2) which has a higher sticking coefficient than tetrahedral white phosphorus (P_4 molecules).

1.1 The Moore's law or the scaling-down law

Integrated circuits (ICs, also known as microchips or microcircuits) have been a major driving force to revolutionize electronic technology in the past few decades. ICs are nowadays used in almost all electronic equipment and devices and are the foundation of the current generation of computers. An IC is a miniaturized electronic circuit consisting of active devices, i.e., transistors and diodes, as well as interconnects elements or passive components (resistance and capacity). Among all these semiconductors, silicon (Si) is absolutely the most important and widely used material in the IC and semiconductor industry owing to its superior properties and low fabrication cost. Before 2005, silicon IC had been developed in an extraordinary pace for almost four decades, known as Moore's law that the number of transistors in an integrated circuit doubles roughly every 18 months [1]. Upon the increase of the number of transistors, each transistor also gets smaller, faster and cheaper. The scalability is the main reason of the tremendous success of many Si IC based technologies, such as Si complimentary metal-oxide-semiconductor (CMOS) technology, which is used to fabricate the central processing unit (CPU) of modern computers. The scalability of Si-CMOS technology is not only about the shrinkage of the dimensions of the devices, but also a number of other factors for maintaining the power density while boosting the performance. For an ideal constant-field scaling [2], upon the shrinkage of all the physical dimensions by α (scaling factor), the depletion depth d also has to be shrunk by α to assure the device works properly. The reduction of the depletion depth requires the increase of doping Nd and the decrease of the applied voltage V by a same factor α since $d \approx \sqrt{N_d V}$ [3]. A direct consequence of the scaling is the increased circuit density by α_2, which dramatically reduces the manufacturing costs. A second important result is the increase of the circuit speed due to both the reduced transit time in transistors and the capacitance in RC delay. However, the applied voltage is found to be impossible to scale by α as continuously shrinking the dimensions because of constraints on the threshold bias in order to avoid rising standby power in the "off" state. Eventually, the applied voltage cannot be scaled anymore, which, unfortunately, already occurred a couple of years ago. This results in the increase of the electric field with the scaling, leading to the increase of the power density of the circuit. It has been shown that the passive power density becomes dominant below the device dimension of 130–65 nm regime [4], which effectively breaks off traditional scaling in CMOS.

In order to further increase the clock frequency without the help of the scaling, the entire architecture should be re-examined. In today's CMOS technology, the intrinsic speed of the transistor is beyond the speed in any other components of the circuit. The intrinsic frequency of a commercial logic circuit transistor is in the order of 102 GHz [5], while other technologies can easily achieve even higher speed, such as SiGe transistors at 500 GHz [6]. Therefore, the speed bottleneck of the ICs results from propagation delay in the passive components, which is dominated by RC (resistance capacitance) delay. Thus, design a new interconnects system in ICs becomes the key to further enhancement of the speed. A couple of approaches have been investigated to replace the predominant aluminum (for electrical conduction) and silicon oxide (SiO_2, for electrical insulation) interconnects including the use of copper and low-k dielectric materials (such as doped SiO_2 or polymeric dielectrics) to reduce the RC product.

Among all these attempts, the optical interconnect design implemented by silicon photonics is extremely promising and can be the potentially ultimate solution to this problem. As shown in **Figure 1**, according to the 2008 International Technology Roadmap for Semiconductors (ITRS) [5], with continued technology node scaling, the relative delay for logic devices and local interconnects decreases. However, global interconnects, especially global interconnects without repeaters, show a dramatic increase in delay time. All of these arguments show that the on-chip interconnects are one of the major challenges for the future IC industry.

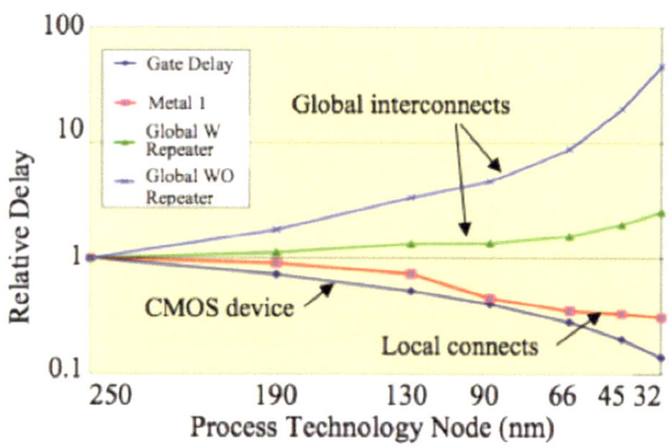

Figure 1.
Relative delay vs. process technology node from 2008 ITRS [5].

1.2 Silicon photonics

1.2.1 Photonics in optical communications

Photonics is about the science and the technology of generating, manipulating, and detecting photons. Silicon, while dominating the semiconductor electronics for decades, is naturally a challenge for developing photonics in many communication applications [7]. Silicon photonics offer a promising platform for the monolithic integration of optics and microelectronics, aiming for many applications including the optical interconnects solution to the microelectronics bottleneck [8]. Because of its importance and rapid growth, a roadmap of silicon photonics, assessed by academic and industrial experts, has been proposed [9]. Light or photon has been used to transmit information for over three decades, mainly by using optical fibers to form optical interconnection between places in distance. In the past three decades, photons have been widely used in optical fiber communication systems, especially in long-distance communications. The fundamental reason for the optical interconnect advantage is the zero rest mass of a photon, which can greatly reduce the required energy. In addition, compared to electrical interconnections, the most important advantage using optical fiber communications is the high speed. However, many challenges need to be solved to achieve on-chip optical interconnects. The first issue is to integrate all the optical devices with the silicon microelectronic devices on silicon chips. Another challenge is materials used in each device. For traditional optical components, III-V materials, such as GaAs and InP, are exclusively used due to their excellent optical properties. However, all the materials and their fabrication are not compatible with the Si exciting Si-CMOS technology.

1.2.2 Key devices for Si photonics

The on-chip interconnect system contains an integrated light source, a modulator to transfer the electrical signal to an optical signal, a waveguide or waveguide device to direct the optical signal to the destination, and finally a photodetector to convert the optical signal back to an electrical signal. Most of these devices are already developed on a Si platform with high bandwidth capability.

Figure 2 Si photonics key devices are needed to realize on-chip optical interconnects: Integrated light source, modulator, waveguide and photo-detector [10]. The only missing key device is an electrically pumped laser source, which is the compatible with the integrated CMOS technology. As we have mentioned above, numerous approaches have been investigated, including porous Si, semiconducting silicides, Er-doped Si, Si nanocrystals or Ge/Si self-assembled quantum dots but none of them could produce a high emission efficiency at room temperature. From the material point of view, in order to achieve a CMOS compatible light source, the

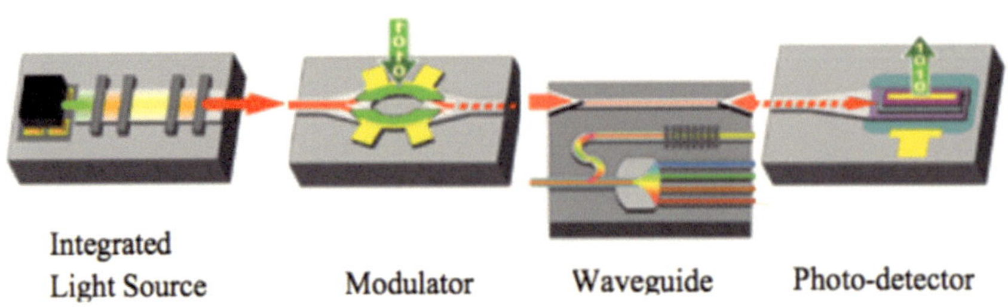

Figure 2.
Illustrates several key photonics devices needed to realize Si on-chip interconnections [10].

use of group IV materials (such as Si, Ge, or Sn) is highly expected. In order to have an efficient light source, a direct band gap semiconductor is preferred. At least, a local minimum at the Γ point of the conduction band is required to accumulate electrons and achieve efficient radiative recombination. This requirement makes Si impossible to be an efficient light source because the difference between the direct and indirect valleys in Si is larger than 2 eV. Fortunately, germanium, which is a group IV material, has a local minimum at the Γ point of the conduction band. More attractively is that the lowest energy point in the Ge conduction band is at the L point, which is only 0.140 eV lower than the lowest energy at the Γ point at room temperature. Therefore, Ge has the potential to be engineered to become a direct band gap material and used as an on-chip integrated light source [11, 12].

1.3 Ge band structure engineering

As previously discussed, Ge is the most interesting group-IV material for the light emitting process. However, achieving direct band gap in Ge and improving the Ge light emitting efficiency are still huge challenges. Indeed, Ge is normally recognized as a poor light emitting material due to its indirect band structure. The radiative recombination through the indirect band-to-band optical transition is inefficient as a result of a phonon-assisted process. The direct band-to-band optical transition in Ge is however a very fast process with a radiative recombination rate of 4–5 orders of magnitude higher than that of the indirect transition [13]. Thus, the direct gap emission

Of Ge can be, in principle, as efficient as that from direct gap III-V materials. The challenge is that the number of the electrons for the direct optical transition is deficient due to an indirect band structure. Fortunately, Ge is a pseudo direct band gap material because of a small energy difference (140 meV) between its direct gap and indirect gap. It has been shown that with a combination of tensile strain and n-type doping Ge can be engineered to be a direct band gap material.

1.3.1 Ge band structure at equilibrium and under injection

The band structure of bulk Ge is shown in **Figure 3**. The valence band is composed of a light-hole band, a heavy-hole band, and a split-off band from spin-orbit interaction. The light-hole band and the heavy-hole band are degenerate at wave vector k = 0 or Γ point, which is the maximum of valence band.

The lowest energy point of the conduction band is located at k = <111> or L point. The energy difference between the conduction band at L point and the valence band at Γ point determines the narrowest band gap in Ge:

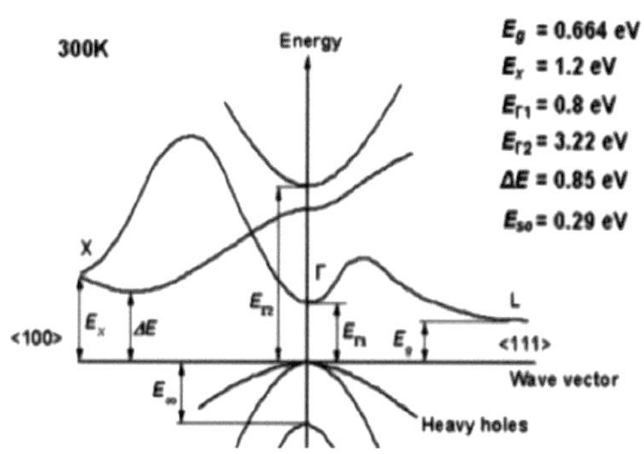

Figure 3.
Ge band structure at 300 K.

Eg = 0.664 eV. Because the direct energy gap $E_{\Gamma 2}$ is much larger than $E_{\Gamma 1}$ and Eg, almost no electrons can occupy such high energy levels. Therefore, we refer direct band gap only to $E_{\Gamma 1}$ throughout this thesis. The part of the conduction band near Γ point is called direct valley and the part near L point is called indirect valley. Since the energy is 4-fold degenerate with regard to the changes of the secondary total angular-momentum quantum number, four L valleys are considered.

Figure 4 shows the electron and the hole distributions of Ge at equilibrium at equilibrium and under injection conditions. At equilibrium, most of the thermally activated electrons occupy the lowest energy states in the indirect L valleys while it is worth noting that in a direct band gap material such as GaAs or InGaAs most of the electrons stay in the direct Γ valley.

Under injection conditions, there are a non-negligible amount of electrons in the Γ valley owing to the small energy difference (140 meV) between the direct band gap and the indirect band gap of Ge, as shown in **Figure 4(b)**. The excess electrons in the Γ valley lead to recombination with the holes in the valence band, which is a highly efficient light emission process because that the direct band-to-band radiative recombination is generally faster than the nonradiative recombinations, such as Auger and defect-assisted processes. But the overall light emission efficiency is very low because most of the injected electrons, staying in the L valleys, recombine nonradiatively due to a slower indirect phonon-assisted radiative recombination than the non-radiative recombinations. On the contrary, the light emission in a direct band gap material such as InGaAs is very efficient because almost all injected electrons are in the Γ valley thus recombine radiatively. To improve the light emission efficiency in Ge, more injected electrons are required to be pumped into Γ valley at the same carrier injection level. Thus, the band structure of Ge can be engineered to accomplish this goal.

1.3.2 Effect of tensile strain on Ge band structure

Efficient direct gap light emission in Ge requires a large amount of electrons in the direct valley. The ratio of the number of the direct valley electrons to the indirect L valleys electrons is determined by the energy difference between the direct band gap and the indirect band gap at quasi-equilibrium. Band structure is associated with crystal lattice, which can be changed by the existence of strain. This effect can be calculated using a strain-modified k·p formalism, which is known to provide an accurate description of the valence and conduction bands all over the Brillouin zone [14]. Such a calculation shows that strain changes the energy levels of the

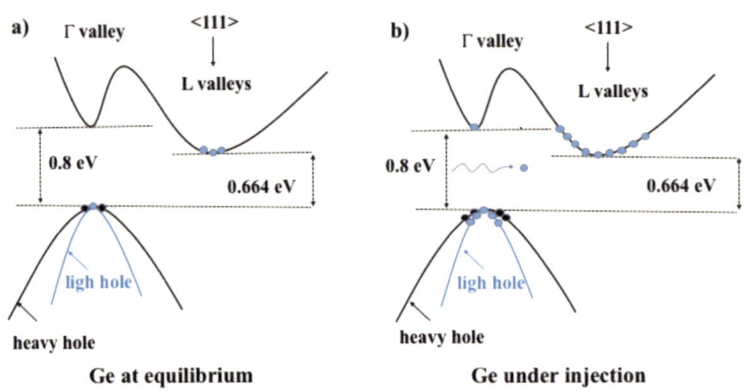

Figure 4.
Distributions of electrons and the holes in Ge at equilibrium and under injection [3].

direct Γ valley, the indirect L valleys the light-hole band, and the heavy-hole band relative to vacuum level. Moreover, the light-hole and the heavy-hole band become nondegenerate and separate at Γ point. **Figure 5** illustrates the variation of direct and the indirect band gaps of Ge under strain [14]. We can see both the direct band gap and the indirect band gap shrink with tensile strain and the direct band gap shrinks faster than the indirect band gap. The direct band gap becomes equal to the indirect band gap at $\varepsilon_{//} \approx 1.8$–1.9% where Ge becomes a direct band gap material. The carrier distribution and the light emission properties of 1.8% tensile-strained Ge under injection are schematically shown in **Figure 6**. Since the Ge becomes a direct gap material, a considerable amount of the excess electrons occupying the direct Γ valley capable of radiative recombination leading to efficient direct gap light emission. The overall emission efficiency of the strained Ge is comparable to that of direct band gap semiconductors.

Biaxial tensile strain is thus an effective way to transform Ge to a direct band gap material, nevertheless two issues exist:

Firstly, highly strained, high-quality single crystalline Ge film is difficult to form because the large lattice change causes thermodynamic instability resulting in dislocations, surface roughness, and other lattice defects. Tensile strain can be induced either by lattice mismatch or by thermal mismatch. The former approach requires a substrate material with larger lattice constant than that of Ge. The latter approach

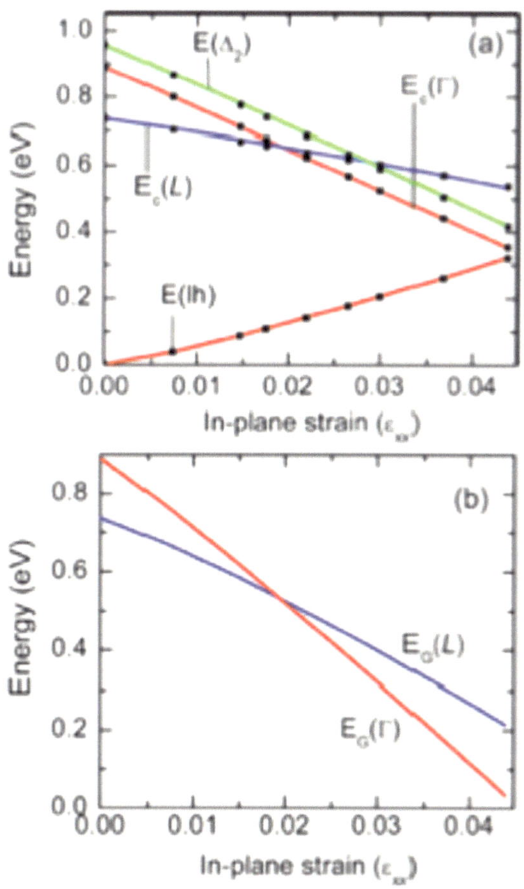

Figure 5.
Energy dependence at low temperature (4 K) of the Γ, Δ2, L, and lh extrema as a function of the in-plane biaxial strain; (b) variations of the direct and indirect band gap energies as a function of the in-plane biaxial strain [14].

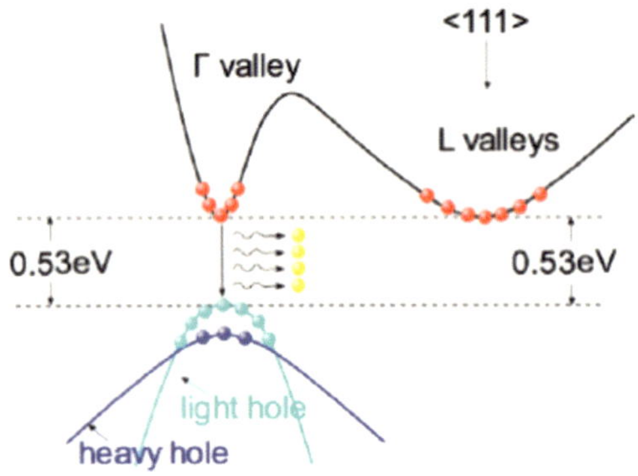

Figure 6.
Electron and hole distributions and light emission at 1.8% tensile-strained Ge.

requires different thermal expansion coefficient between Ge and the substrate material, which is adopted in this research. Up to 0.25–0.30% tensile strain has been achieved in Ge epitaxially grown on Si substrate, which will be discussed later. The second issue is the excessive change of the band gap in highly tensile-strained in Ge. Both the direct band gap and the indirect band gap become 0.53 eV at 1.8% tensile strain, as shown in **Figure 6**. This optical band gap is corresponding to an emission wavelength of about 2300 nm, which is far away from the 1550 nm telecommunication wavelength band, which is also the primary choice for Si photonics. These two issues suggest that very high tensile strains are not favorable in both material growth and photonics applications. Thus, the increase of the number of the injected excess electrons in the Γ valley owing to strain effect is limited. This problem can be solved by n-type doping in Ge.

1.4 Growth of tensile-strained Germanium on silicon substrates

1.4.1 Ge band gap engineering induced by tensile strain

Semiconductor diode lasers are conventionally based on direct band gap materials due to the efficient radiative recombination of direct gap transitions. As discrete devices, direct band gap III-V semiconductors are the main materials used to fabricate semiconductor lasers in telecommunications. On the other hand, indirect semiconductors, such as Si, Ge and their SiGe alloys are traditionally considered unsuitable for laser diodes due to their indirect band structure. The radiative recombination through the indirect band to-band optical transition is inefficient as a result of a phonon-assisted process. However, the direct band-to-band optical transition in Ge was shown to be very fast, exhibiting radiative recombination rate of 4–5 orders of magnitude higher than that of the indirect transition [13, 15]. Thus, if we can highly increase the number of electrons in the direct band gap, Ge could be used to fabricate diode lasers. Among group-IV indirect semiconductors (Si, and SiC), Ge exhibits interesting properties both from the transport and optical points of view. Regarding the transport properties, as can be seen from the **Table 1**, the hole mobility of Ge is highest in all semiconductors and its electron mobility is about 2.7 times higher than that of Si. Thus, Ge is of great interest for high mobility CMOS transistors, in particular p-CMOS transistors. Concerning the optical properties, Ge can be considered as a pseudo direct band gap material because of a small difference in energy between its direct gap and indirect gap, which is 0.140 eV or ~ 5 kBT at room temperature (kB is the Boltzmann constant). Theoretically, it has been shown that Ge can be transformed into a direct gap material with ~1.8–1.9% tensile strain

Properties	Si	Ge	GaAs	InAs	GaP	InP	InSb
E_g [eV]	1.124	0.664	1.424	0.354	2.272	1.344	0.225
μ_n [cm²/V.s]	1450	3900	8000	30.000	200	5000	80.000
μ_p [cm²/V.s]	370	1800	400	480	150	180	1500

Table 1.
Values of the forbidden band gap and mobilities of holes and electrons in group-IV and III-V semiconductors.

[14, 16–18]. Indeed, upon application of biaxial tensile strain, both direct and indirect gaps reduce but the direct gap reduces faster. Thus, Ge can progressively transform from an indirect gap semiconductor toward a direct gap semiconductor with the increase of tensile strain. Furthermore, the tensile strain also lifts the degeneracy between heavy hole and light hole valence bands. The small effective mass of the light hole band reduces the density of states in the valence band, which, in turn, decreases the threshold for optical transparency and lasing [14, 18].

Tensile strain can be induced in Ge via several approaches: applying external mechanical stress [19], growing Ge on a larger lattice parameter substrate, such as InGaAs/GaAs [20, 21] or on relaxed GeSn buffer layers on Si [22, 23], or by taking benefit of the thermal mismatch between Ge and Si [18, 24–26]. An interesting result recently demonstrated by Jain et al. [27], who realized a Ge-MEMS device with 1% tensile strain introduced by a Si_3N_4 stressor. Compared to bulk Ge, the authors observed an enhancement of the photoluminescence intensity up to 260 times. In the frame of this work, we focus on the effect of band gap engineering of Ge films directly grown on Si substrates. The main advantage of the Ge/Si system is that it allows direct integration of Ge-based optoelectronic devices into the mainstream Si technology. In view of device applications, it is vital to obtain high-quality epitaxial material growth, i.e., to get Ge epilayers, which have a smooth surface and a low density of threading dislocations. To reach such an objective, the challenge is to control the Ge/Si growth process to overcome the limitation imposed by 4.2% lattice mismatch between these two materials. Thus, in the following part, we present results on the study of Ge growth on Si using a two-step method and on the dependence of the strain level on the substrate temperature.

1.4.2 Effect of the growth temperature on the value of tensile strain

After the low-temperature growth step, we have deposited another Ge film at higher substrate temperatures. In the following, we will consider the effect of the two-step growth process on the tensile strain in the Ge films. All samples have a total thickness of ~ 300 nm, which consists of a ~50 nm thick Ge film deposited at 300°C followed by a 250 nm thick Ge layer grown at various temperatures: 400, 500, 600, 650, 700, 750 and 770°C. As the tensile strain in Ge layers is induced by the difference of the thermal expansion coefficients between Ge and Si, it is natural to expect that the level of tensile strain should increase with increasing the growth temperature. **Figure 7** displays some representative θ–2θ XRD scans taken around the Ge(004) reflection. For comparison, we report in dotted lines a XRD scan of a sample grown at 300°C of the same thickness.

As the growth temperature increases, the Ge(004) peak linearly shifts to higher angles, reaches a saturation value at 700°C and finally remains almost constant for further increasing the temperature to 770°C. The in-plane tensile strain observed in the temperature range of 700–770°C is 0.24%. The fact that the (004) peak of the 300°C grown sample is located at 2θ ~ 66°, a value close to that measured on a Ge substrate, indicates that the corresponding Ge layer is almost fully relaxed. By using a high-resolution XRD, we estimate the rate of strain relaxation is about 95–96%. As can be seen in **Figure** 7, the 2θ angle value of the (004) peak is found to linearly increase when the growth temperature increases from 300 to 700°C then remains almost constant for further increasing the temperature up to 770°C. We note that to obtain the value of the in-plane tensile strain ε//, we first determine the out-of plane strain $\varepsilon_\perp$ from the θ–2θ XRD curves and then deduce the value of ε// using the following relationship: $\varepsilon_{//}/(\varepsilon_{//} + \varepsilon_\perp) = c_{11}/(c_{11} + 2c_{12})$ with $c_{11} = 12.85 \times 10^{10}$ Pa and $c_{12} = 4.83 \times 10^{10}$ Pa for pure Ge [28]. The highest value of the in-plane tensile strain $\varepsilon_{//}$ obtained in the growth temperature range of 700–770°C is 0.24%, which is in agreement with previous results reported using CVD in which the highest tensile strain was in the range of 0.22–0.25% [24–26, 29–33].

As we have already mentioned previous part, one of the unique properties of Ge is the very small energy difference between its direct and indirect band gap. Thus, Si/Ge interdiffusion represents an obstacle to overcome in order to develop Ge-based optoelectronic devices. Si/Ge interdiffusion has been shown to greatly affect the optical properties of Si/Ge heterostructures [34] and degrade the performance of metal-oxide semiconductor field-effect transistors (MOSFET) by reducing strain and carrier confinement and increasing alloy scattering [35]. To prevent interdiffusion or out-diffusion of an element (occurs when the sample was annealed at high temperature to enhance the tensile strain value in the Ge film), it is common to use a diffusion barrier and materials must be not only nonreactive but also are able to strongly adhere to adjacent materials. In electronic or memory devices, multilayers of metals, WN_2, RuTiN, or RuTiO, are usually used to prevent out diffusion of dopants (B and P) or oxidation of devices [36–38]. Such materials are, however, difficult insert in a heterostructure where epitaxial growth is needed. It is now generally accepted that Si and Ge atoms interdiffuse via both vacancy and interstitial-related mechanisms [39]. To prevent Si/Ge interdiffusion, in particular to suppress upward diffusion of Si to the deposited Ge layer during growth or subsequent annealing, we have experimented an approach to saturate vacancies and

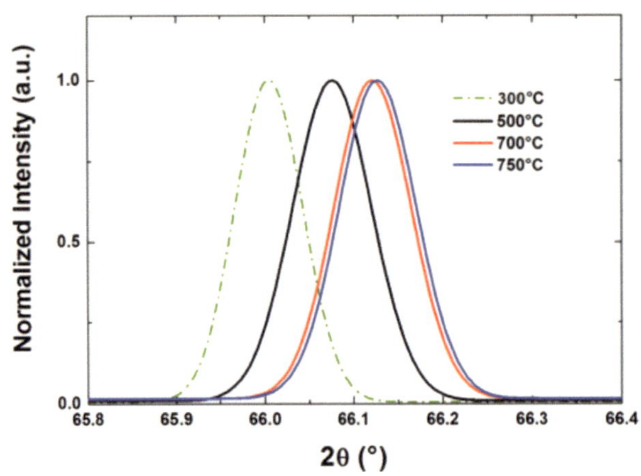

Figure 7.
Evolution of θ–2θ XRD scans around the Ge(004) reflection with the growth temperature. The doted XRD scan corresponding to a sample grown at 300°C is shown for comparison.

interstitial sites in the Ge layer near the interface region. We have chosen carbon for its small atomic radius (twice as small as Ge), carbon atoms are thus highly mobile and can easily diffuse via interstitial mechanism [40–42].

To further confirm the effect of Ge/Si diffusion, we have used high-resolution HR-XRD reciprocal space mapping (RSM) equipped with a rotating anode to determine the Si composition and the strain level in Ge layers. 004, 224, and −2−24 RSMs were collected along the four <110> azimuthal directions. Combining the (224) map with that along (004) direction, one can measure both parallel and perpendicular lattice parameters of the Ge layer. The data obtained in reciprocal space are converted in direct space. We show in **Figure 8** the values of inter-reticular distances of the Si substrate and the Ge layer, measured along the (004) and (220) planes, of the sample annealed by 10 cycles at 780/900°C for 20 min. Note that for the as-grown sample, we obtain $a_{//}$ = 5.669 Å and $a_{\perp}$ = 5.649 Å, which correspond the value of pure Ge (a_{Ge} = 5.657 Å). For the annealed sample, we obtain $a_{//}$ = 5.656 Å and $a_{\perp}$ = 5.625 Å. Thus, the volume occupied by this unit cell corresponds to a Si_xGe_{1-x} alloy with a Si average concentration x = 5% (the corresponding average lattice parameter is about 5.646 Å).

Figure 9(a) shows a TEM image taken near the interface region of a sample in which we have inserted three separate carbon layers during the first step of Ge growth. Starting from a clean and (2 × 1) reconstructed Si surface, we first grow a 30 nm thick Ge buffer layer at 300°C and then about 4 monolayers (ML) of carbon at the same temperature. Since epitaxial growth of the upper Ge layers should be conserved, the amount of deposited carbon is crucial. The latter can be determined using the change of RHEED patterns. Upon carbon deposition at 300°C, the (2 × 1) RHEED pattern characteristic of a clean Ge(001) surface remains almost unchanged up to carbon coverage of 6 MLs, beyond which a faint pattern appears. Since it is crucial to insure epitaxial growth of the upper Ge layer, a carbon amount of 4 MLs is then chosen (the corresponding carbon thickness is ~0.3 nm). After this step, a Ge layer is deposited on top of carbon, producing therefore C/Ge stacked layers in order to increase the efficiency of carbon-induced diffusion barriers. We have experimented three multilayers of C (0.3 nm)/Ge (18 nm). **Figure 9(b)** shows

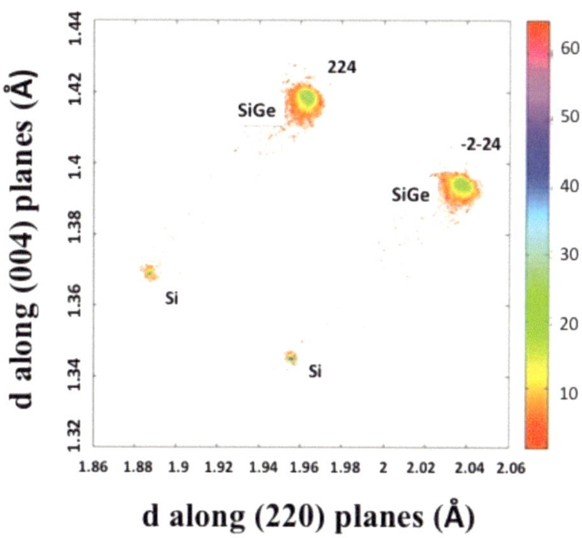

Figure 8.
Values of inter-reticular distances of Si and Ge, measured along the (004) and (220) planes of a 300 nm thick Ge layer deposited on Si at 730°C, followed by 10 cyclic anneals at 780/900°C for 20 min.

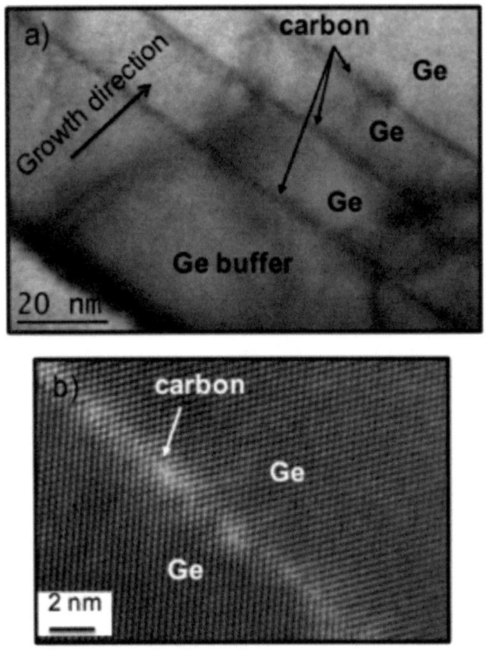

Figure 9.
(a) TEM image taken near the interface region illustrating the growth of three multilayers of C (0.3 nm)/Ge (18 nm); (b) atomically resolved TEM image taken in the vicinity of the carbon layer.

an atomically resolved TEM image taken in the vicinity of the carbon layer. Clearly, the underneath Ge layers and also the upper Ge layers are perfectly epitaxial, without any detectable defects. The image also reveals that carbon atoms are distributed over a distance of ~2 nm, probably to occupy the interstitial sites of the Ge lattice or due to the strain generation arising from carbon insertion.

To verify the efficiency of C/Ge multilayers to prevent Si/Ge interdiffusion, we have finally grown a 300 nm thick Ge layer following the two-step growth as described previously. The high temperature step was carried out at 730°C. After growth, the sample was cyclically annealed by 10 cycles from 780 to 900°C and the annealing time was 20 min. In **Figure 10**, we display SIMS measurements of two samples: one without carbon deposition (red curve) and the other containing three C/Ge multilayers. As can be seen, the as-grown sample (black) shows a relatively sharp interface, no long-range upward diffusion of Si from the substrate was observed and the deposited Ge layer remains pure. On the other hand, the Si profile of the annealed sample, indicated in blue, reveals a pronounced Si upward diffusion. The Si content within the Ge film is highly heterogeneous, it continuously decreases from the interface to the film surface. The average Si content in the Ge film is estimated to be about 5%.

1.5 Phosphorus doping using a GaP decomposition source

Heavy n-doping of Ge films is essential to achieve efficient light emission from the direct gap transition. However, heavy n-type doping in Ge is a challenge due to a low solubility and a fast diffusivity of dopants. Three elements: phosphorus, arsenic and antimony, can be used as n-type doping in group-IV semiconductors. We report in **Table 2** the solubility of these three elements in Ge [43].

It can be seen that phosphorous has the highest solubility and at a temperature not too high. In UHV-CVD, the highest phosphorus concentration of 1.2×10^{19} cm^{-3}

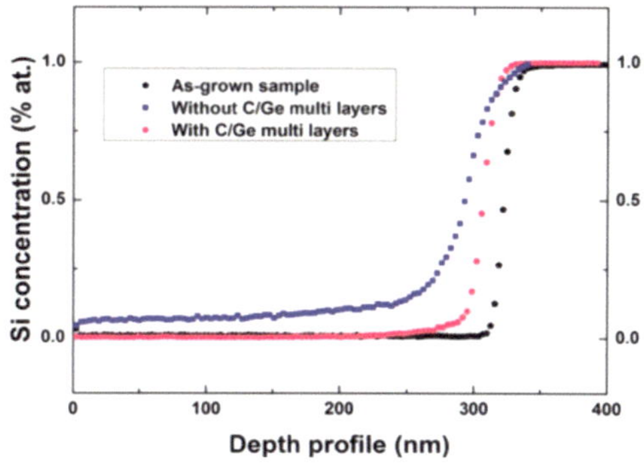

Figure 10.
SIMS profiles measuring the Si content of the as-grown sample (black) and after 10 cyclic anneals at 780/900°C for 20 min (blue). The red curve corresponds to a sample in which three short C/Ge multilayers were deposited near the interface region.

Dopant	Maximum solubility	Temperature (°C)
P	2×10^{20} cm^{-3}	580
As	8.7×10^{19} cm^{-3}	800
Sb	1.1×10^{19} cm^{-3}	800

Table 2.
Solubility of different n-type dopants in Ge [43].

was reported [44]. In CVD, the PH_3 is commonly used as a gas precursor and an activated P concentration of 2×10^{19} cm^{-3} was obtained at a temperatures of 600–700°C [12]. A phosphorus concentration up to 4.5×10^{19} cm^{-3} can be achieved by combining delta doping with diffusion barriers to reduce the P out-diffusion [45, 46]. It is worth noting that while dopant implantation is commonly used in the CMOS process, the activated P concentration is limited at about 1×10^{19} cm^{-3} even with the use of diffusion barriers [47], and a higher P concentration seriously deteriorates the material quality. Since tetrahedral white phosphorus (P_4 molecules) is not stable and highly volatile, we have used a specific doping cell based on the decomposition of GaP to produce di-phosphorus (P_2 molecules) [46–48], which has a sticking coefficient of about 10 times larger than that of tetrahedral white phosphorus. Thus, the phosphorus concentration in the Ge lattice could be increased and we can expect to enhance the efficiency of radiative recombination in Ge.

1.5.1 Specific GaP decomposition source

A photograph of a GaP decomposition cell is shown in **Figure 11**, which is heated by a Ta wire filament supported by PBN rings, similar to standard effusion cells. Compared to a valved phosphorus thermal cracker, the GaP cell is easy to set up, it operates as any effusion cell and is compatible with MBE environments. Using a Ga-trapping cap system, the GaP source can provide high-purity P_2 beam by decomposition of GaP. Its operation is based on the sublimation of phosphorus from GaP at an intermediate temperature range in which Ga has a very low vapor pressure. By placing a cap with small apertures on top of the cell (**Figure 11(b)**),

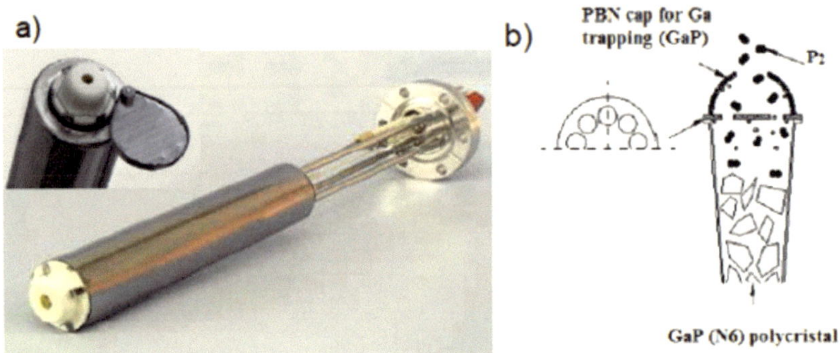

Figure 11.
(a) Photograph of a GaP decomposition cell and (b) schema of the Ga trapping cap [50].

Ga atoms are trapped by the cap and only the P_2 beam can escape. According to the supplier [49], a P_2/P_4 ratio of about 150:1 can be achieved.

1.5.2 Efficiency of phosphorus doping from a GaP source

Figure 12 displays the evolution of the room-temperature photoluminescence spectrum versus the temperature of the GaP cell. For all samples, the substrate temperature is chosen to be 300°C and the film thickness is 100 nm. The temperature of the GaP source increases from 600 to 750°C. Note that based on the ref. [51], we avoid doping Ge films at 800°C. After growth, all samples were annealed in the growth chamber at 750°C during 1 min to activate dopants. As can be seen from the figure, the photoluminescence intensity increases with increasing the temperature of the GaP source temperature from 600 to 725°C and the highest PL intensity is obtained at 725°C. For the GaP source temperature at 750°C, the PL is found to decrease. Thus, the PL result, indicating that above 725°C the Ga trap from the GaP cell becomes less efficient.

1.5.3 Dependence of the substrate temperature on the doping level

To investigate the effect of the doping level versus the substrate temperature, we have therefore kept the GaP source at a constant temperature of 725°C. We display in **Figure 13** the evolution of the photoluminescence spectrum with the

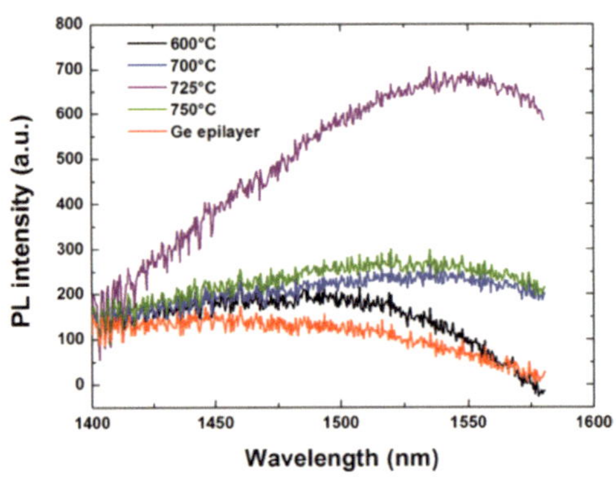

Figure 12.
Evolution of photoluminescence spectrum versus the GaP source temperature. All PL measurements were carried out at room temperature.

substrate temperature. The reference sample is a 600 nm thick undoped Ge layer deposited at a substrate temperature of 170°C. The temperature of the GaP source is 725°C. The spectrum of the reference sample exhibits a very weak intensity and the emission from direct band gap is not clearly observed, as expected in an indirect band gap semiconductor. The PL intensity, which is very weak for the sample grown at 300°C, is found to increase with decreasing the substrate temperature and the highest PL intensity is obtained for a substrate temperature of 170°C. The evolution of the PL signal follows almost the same trend already observed for the electrical measurements not show here and thus confirms a high efficiency of P doping at a low substrate temperature of 170°C. The 170°C PL spectrum peaks at around 1624 nm (the corresponding energy is 0.765 eV). This transition can be attributed to arise from the direct band gap radiative recombination in the n-doped Ge layer. It is worth noting that if we compare the photoluminescence intensity of the sample doped at 170°C (blue curve) with that of the undoped sample (black curve), an intensity enhancement of about 50 times is obtained. Another noteworthy point is that when comparing the energy maximum at around 0.810 eV, arising from the direct band gap emission of unstrained and undoped Ge, we observe here a redshift of 45 meV, which can be attributed to band gap narrowing at high n-doping levels [52]. Indeed, a low temperature growth at 170°C does not induce tensile strain and after annealing at 750°C for 1 min, the corresponding Ge films only exhibit a tensile strain as low as 0.13%. Thus, the above optical redshift can be directly correlated to the effect of n-doping. According to ref. [52], a redshift of 45 meV corresponds to a doping level of $\sim 2 \times 10^{19}$ cm^{-3}. The separate investigation of the effect of the substrate temperature on the tensile strain and n-doping of the Ge layer brings fruitful information. The substrate temperature is shown to produce an opposite effect on these two properties of the Ge film. Higher growth temperatures (up to 770°C) induce larger tensile strain while low temperatures favor n doping. It is noteworthy that an activated n-doping level higher than 10^{19} cm^{-3} can be obtained at a growth temperature of 170°C. In addition, it should be pointed out that if Ge is directly deposited on Si substrates at a substrate temperature of 170°C, the Ge/Si growth does proceed via the Stranski-Krastanov mode. However, once a smooth Ge buffer has been formed on Si at 300°C, further Ge deposition on this Ge buffer layer will behave similarly as a homoepitaxial Ge growth and the growth is two-dimensional in a very large range of substrate temperatures.

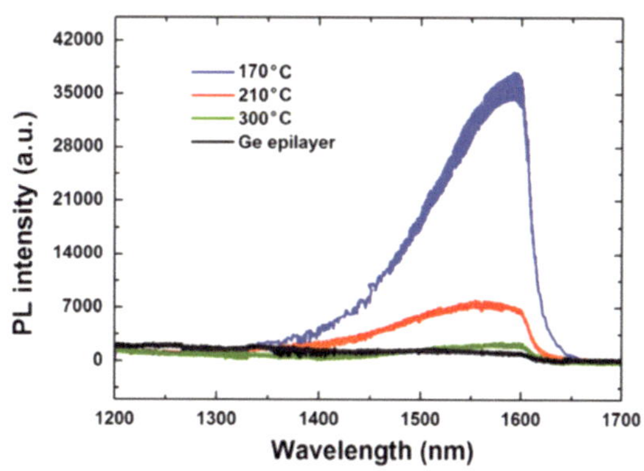

Figure 13.
Evolution of the photoluminescence spectrum versus the growth temperature.

2. Conclusion

To adopt an adequate strategy to produce Ge epilayers for optoelectronic applications, first it is important to emphasize that the thermal-induced tensile train is limited to a value lying in the range of 0.25–0.30%, which is far from the expected value to get direct band gap Ge. Thus, it would be preferable to set up growth conditions, which favor n-doping. In other words, heavy n-doping appears to be the parameter that is more important than the value of the tensile strain. This implies that an adequate growth strategy would be to grow and to n-dope Ge films at low temperatures followed by post-growth thermal annealing. Since, Ge films, which are grown and n-doped at low temperatures, usually contain a high density of point defects (vacancies and interstitials). Therefore, the primary role of post thermal annealing is to restore the crystalline quality and to activate dopants into substitutional sites, without taking care of the value of tensile strain in the film. The efficiency of phosphorus doping may depend on two main parameters: the dopant solubility in a matrix and the sticking coefficient of dopants on the film surface. These two parameters are probably competing. The sticking coefficient of an atom or a molecule on a substrate surface increases with decreasing the substrate temperature. Since our results reveal that P doping is more favorable at low substrate temperatures, it appears that the sticking coefficient of the P_2 molecules is the dominant parameter determining the phosphorus doping level in Ge.

Acknowledgements

We thank Prof. V. Le Thanh, Prof. P. Boucaud, Dr. A. Ghrib and their group for supporting this work.

Author details

Luong Thi Kim Phuong
Hong Duc University, Thanh Hoa City, Vietnam

*Address all correspondence to: luongthikimphuong@hdu.edu.vn

References

[1] Moore G. Cramming more components onto integrated circuits. Electronics. 1965;**38**:144

[2] Dennard RH, Gaensslen FH, Yu H-N, Rideout VL, Bassous E, LeBlanc A. Design of ion-implanted MOSFETs with very small physical dimensions. IEEE Journal of Solid-State Circuits. 1974;**9**:256

[3] Sun W. PhD thesis; Massachusetts Institute of Technology; 2009

[4] Haensch W, Nowak EJ, Dennard RH, Solomon PM, Bryant A, Dokumaci OH, et al. Silicon CMOS devices beyond scaling. IBM Journal of Research and Development. 2006;**50**:339

[5] International Technology Roadmap for Semiconductors; 2008. Available from: http://www.itrs.net/home.html

[6] Krithivasan R, Lu Y, Cressler JD, Rieh J-S, Khater M, Ahlgren D, et al. Half-TeraHertz operation of SiGe HBTs. IEEE Electron Device Letters. 2006;**27**:567

[7] Jalali B, Fathpour S. Silicon photonics. IEEE Journal of Lightwave Technology. 2006;**24**:4600

[8] Haurylau M, Chen G, Chen H, Zhang J, Nelson NA, Albonesi DH, et al. On-chip optical interconnect roadmap: Challenges and critical directions. IEEE Journal of Selected Topics in Quantum Electronics. 2006;**12**:1699

[9] Kirchain R, Kimerling LC. A roadmap for nanophotonics. Nature Photonics. 2007;**1**:303

[10] Available from: www.intel.com

[11] Liu J, Sun X, Pan D, Wang X, Kimerling LC, Koch TL, et al. Tensile-strained, n-type Ge as a gain medium for monolithic laser integration on Si. Optics Express. 2007;**15**:11272

[12] Sun X, Liu J, Kimerling LC, Michel J. Room-temperature direct bandgap electroluminesence from Ge-on-Si light-emitting diodes. Optics Letters. 2009;**34**:1198

[13] Grzybowski G, Roucka R, Mathews J, Jiang L, Beeler RT, Kouvetakis J, et al. Direct versus indirect optical recombination in Ge films grown on Si substrates. Physical Review B. 2011;**84**:205307

[14] El Kurdi M, Fishman G, Sauvage S, Boucaud P. Band structure and optical gain of tensile-strained germanium based on a 30 band k·p formalism. Journal of Applied Physics. 2010;**107**:013710

[15] Haynes JR, Nilsson NG. The direct radiative transitions in germanium and their use in the analysis of lifetime. In: Proceedings of VIIth International Conference on Physics of Semiconductors; Paris. 1964. p. 21

[16] Soref RA, Friedman L. Direct gap Ge/GeSn/Si and GeSn/Ge/Si heterostructures. Superlattices and Microstructures. 1993;**14**:189

[17] Soref RA. Silicon-based optoelectronics. Proceedings of the IEEE. 1993;**81**:1687

[18] Liu J, Sun X, Pan D, Wang X, Kimerling LC, Koch TL, et al. Tensile strained n-type Ge as a gain medium for monolithic laser intergration on Si. Optics Express. 2007;**15**:11272

[19] El Kurdi M, Bertin H, Martincic E, de Kersauson M, Fishman G, Sauvage S, et al. Control of direct band gap emission of bulk germanium by mechanical tensile strain. Applied Physics Letters. 2010;**96**:041909

[20] Bai Y, Lee KE, Cheng C, Lee ML, Fitzgerald EA. Growth of highly tensile

strained Ge on relaxed $InxGa_{1-x}As$ by metal- organic chemical vapor deposition. Journal of Applied Physics. 2008;**104**:084518

[21] Jakomin R, de Kersauson M, El Kurdi M, Largeau L, Mauguin O, Beaudoin G, et al. High quality tensile-strained n-doped germanium thin films grown on InGaAs buffer layers by metal-organic chemical vapor deposition. Applied Physics Letters. 2011;**98**:091901

[22] Fang YY, Tolle J, Roucka R, Chizmeshya AVG, Kouvetakis J, D'Costa VR, et al. Perfectly tetragonal, tensile-strained Ge on $Ge_{1-y}Sn_y$ buffered Si (100). Applied Physics Letters. 2007;**90**:061915; Menëndez J, Kouvetakis J, ibid. 2004;**85**:1175

[23] Takeuchi S, Shimura Y, Nakatsuka O, Zaima S, Ogawa M, Sakai A. Growth of highly strain-relaxed $Ge_{1-x}Sn_x$/virtual Ge by a Sn precipitation controlled compositionally step-graded method. Applied Physics Letters. 2008;**92**:231916

[24] Liu J, Sun X, Aguilera RC, Kimerling LC, Michel J. Ge-on-Si laser operating at room temperature. Optics Letters. 2010;**35**:679

[25] Ishikawa Y, Wada K. Gemanium for silicon photonics. Thin Solid Films. 2010;**518**:S83

[26] Liu J, Camacho-Aguilera R, Bessette JT, Sun X, Wang X, Cai Y, et al. Ge-on-Si optoelectronics. Thin Solid Films. 2012;**520**:3354

[27] Jain JR, Hryciw A, Baer TM, Miller DAB, Brongersma ML, Howe RT. A micromachining-based technology for enhancing germanium light emission via tensile strain. Nature Photonics. 2012;**6**:398

[28] Liu J, Cannon DD, Ishikawa Y, Wada K, Danielson DT, Jongthammanurak S, et al. Deformation potential constants of biaxially tensile strained Ge epitaxial films on Si(100). Physical Review B. 2004;**70**:155309

[29] Colace L, Masini G, Galluzzi F, Assanto G, Capellini G, Di Gaspare L, et al. Metal-semiconductor-metal near-infrared light detector based on epitaxial Ge/Si. Applied Physics Letters. 1998;**72**:3175

[30] Luan HC, Lim DR, Lee KK, Chen KM, Sandland JG, Wada K, et al. High-quality Ge epilayers on Si with low threading- dislocation densities. Applied Physics Letters. 1999;**75**:2909

[31] Hartmann J-M, Abbadie A, Papon A-M, Holliger P, Rolland G, Billon T, et al. Reduced pressure-chemical vapor deposition of Ge thick layers on Si(001) for 1.3-1.55-μm photodetection. Journal of Applied Physics. 2004;**95**:5905

[32] Hartmann J-M, Papon A-M, Destefanis V, Billon T. Reduced pressure chemical vapor deposition of Ge thick layers on Si(001), Si(011) and Si(111). Journal of Crystal Growth. 2008;**310**:5287

[33] Halbwax M, Bouchier D, Yam V, Débarre D, Nguyen LH, Zheng Y, et al. Kinetics of Ge growth at low temperature on Si(001) by ultrahigh vacuum chemical vapor deposition. Journal of Applied Physics. 2005;**97**:064907

[34] Boucaud P, Wu L, Guedj C, Julien FH, Sagnes I, Campidelli Y, et al. Photoluminescence and intersubband absorption spectroscopy of interdiffused Si/SiGe quantum wells. Journal of Applied Physics. 1996;**80**:1414

[35] Jung J, Yu SF, Olubuyide OO, Hoyt JL, Antoniadis DA, Lee ML, et al. Effect of thermal processing on mobility in strained Si/strained SiGe on relaxed SiGe virtual substrates. Applied Physics Letters. 2004;**84**:3319

[36] Yoon DS, Roh JS, Lee SM, Baik HK. Alteration for a diffusion barrier design concept in future high-density dynamic and ferroelectric random access memory devices. Progress in Materials Science. 2003;**48**:275

[37] Becker S, Gordon RG. Diffusion barrier properties of tungsten nitride films grown by atomic layer deposition from bis(tert-butylimido) bis(dimethylamido) tungsten and ammonia. Applied Physics Letters. 2003;**82**:2239

[38] Ting CY. TiN as a high temperature diffusion barrier for arsenic and boron. Thin Solid Films. 1984;**119**:11

[39] Dong Y, Lin Y, Li S, McCoy S, Xia G. A unified interdiffusivity model and model verification for tensile and relaxed SiGe interdiffusion over the full germanium content range. Journal of Applied Physics. 2012;**111**:044909

[40] Spiesser A, Slipukhina I, Dau MT, Arras E, Le Thanh V, Michez L, et al. Control of magnetic properties of epitaxial $Mn_5Ge_3C_x$ films induced by carbon doping. Physical Review B. 2011;**84**(165203)

[41] Dau MT, Le Thanh V, Le TG, Spiesser A, Petit M, Michez LA, et al. Mn segregation in Ge/Mn(5)Ge(3) heterostructures: The role of surface carbon adsorption. Applied Physics Letters. 2011;**99**:151908

[42] Dau MT, Le Thanh V, Le TG, Spiesser A, Petit M, Michez LA, et al. Suppression of Mn segregation in Ge/ Mn_5Ge_3 heterostructures induced by interstitial carbon. Thin Solid Films. 2012;**520**:3410

[43] Madelung O. Physics of Group IV Elements and III-V Compounds, Landolt-Börnstein: Numerical Data and Functional Relationships in Science and Technology. Vol. 17a. Berlin: Springer; 1982

[44] Sun X, Liu J, Kimerling LC, Michel J. Optical bleaching of thin film Ge on Si. ECS Transactions. 2008;**16**:881

[45] Camacho-Aguilera R, Cai Y, Bessette JT, Kimerling LC, Michel J. High active carrier concentration in N-type, thin film Ge using delta-doping. Optical Materials Express. 2012;**2**:146211

[46] Camacho-Aguilera R, Han Z, Cai Y, Kimerling LC, Michel J. Direct band gap narrowing in highly doped Ge. Applied Physics Letters. 2013;**102**:152106

[47] Ding L, Lim AEJ, Liow JTY, Yu MB, Lo G-Q. Dependences of photoluminescence from P-implanted epitaxial Ge. Optics Express. 2012;**20**:8228

[48] Friess E, Nutzel J, Abstreiter G. Phosphorous doping in low temperature silicon molecular beam epitaxy. Applied Physics Letters. 1992;**60**:2237

[49] Shitara T, Ebert K. Electronic properties of InGaP grown by solid-source molecular-beam epitaxy with a GaP decomposition source. Applied Physics Letters. 1994;**65**:356

[50] Available from: http://www.mbe-komponenten.de

[51] Luong TKP et al. Molecular-beam epitaxial growth of tensile-strained and n-doped Ge/Si (001) films using a GaP decomposition source. Thin Solid Films. 2014;**557**:70

[52] Oehme M, Gollhofer M, Widmann D, Schmid M, Kaschel M, Kasper E, et al. Direct bandgap narrowing in Ge LED's on Si substrates. Optics Express. 2013;**21**:2206

Chapter 6

Double-Sided Passivated Contacts for Solar Cell Applications: An Industrially Viable Approach Toward 24% Efficient Large Area Silicon Solar Cells

Zhi Peng Ling, Zheng Xin, Puqun Wang, Ranjani Sridharan, Cangming Ke and Rolf Stangl

Abstract

Tunnel layer passivated contacts have been successfully demonstrated for next-generation silicon solar cell concepts, achieving improved device performance stemming from the significantly reduced contact recombination of the solar cell contacts. However, these carrier-selective passivated contacts are currently deployed only at the rear side of the silicon solar cell, while the front side adopts a conventional diffused junction and contacting scheme. In this work, we report on the novelty and feasibility of deploying tunnel layer passivated contacts on both sides of a silicon wafer-based solar cell, featuring a textured front surface and a planar rear surface. In particular, we demonstrate that silicon solar cells incorporating our in-house developed electron-selective thermal-SiO_x/poly-Si(n^+) and hole-selective thermal-SiO_x/poly-Si(p^+) passivated contacts have a solar cell efficiency potential approaching 24%. Deploying contact passivation only at the rear side of the solar cell, we have reached a solar cell efficiency of 21.7%, using conventional screen printing for metallization. We present a systematic approach of evaluating our in-house developed electron-selective and hole-selective passivated contacts on both textured and planar lifetime test structures, as well as dark I–V test structures, to extract the recombination current density j_0 and the contact resistance R_c of the passivated contact, which is used for process optimization as well as for subsequent efficiency potential prediction. The two key challenges aiming at a double-sided integration of passivated contacts are (1) parasitic absorption within the front-side highly doped poly-Si capping layer, requiring the processing of ultrathin (≤10-nm) contact passivation layers. This has been quantified by numerical simulation (using SunSolve™) and also solved experimentally, i.e., processing ultrathin 3-/10-nm hole/electron extracting SiO_x/poly-Si(p^+/n^+) passivated contact layers, reaching an implied open-circuit voltage of 690/703 mV on a planar/textured silicon surface, which will even further enhance after SiN_x capping. (2) Ensuring process compatibility with conventional screen printing: Screen printing on electron extracting poly-Si(n^+) seems feasible; however, screen printing on hole-extracting poly-Si(p^+) is still a challenge. Solar cell precursors have been processed, showing excellent pre-metallization results (implied-V_{OC} ~ 713 mV). According to our efficiency potential prediction (using the measured j_0 and R_c values of our developed contact passivation layers), bifacial double-sided passivated contact solar cells (efficiency potential of ~23.2%, using our layers) can clearly outperform rear-side-only passivated contact solar cells (efficiency potential of ~22.5%).

Keywords: passivated contacts, contact passivation, silicon solar cells, double-sided passivated contacts

1. Introduction

To meet the future energy needs, there is a need to develop low-cost alternative energy sources to complement the conventional energy sources (e.g., oil, gas, coal) as well as to address the pressing environmental issues associated with the latter. Hence, energy-related technology roadmaps are actively being released and revised toward the future energy needs. One good example is the International Technology Roadmap for Photovoltaic (ITRPV) [1]. In general, a successful deployment of any selected solar cell technology will be mainly dominated by (i) cost-effectiveness of the material and processes, (ii) scalability to high-volume manufacturing, (iii) device performance, and (iv) long-term stability of product. To progress toward item (iii), ITRPV predicts a continuous reduction of recombination losses in the wafer as well as at the front and rear surfaces of the solar cell. According to Ref. [2–4], given the considerable improvements in the wafer bulk, and surface passivation layers, the main source of recombination losses in high-efficiency solar cells is now dominated by the metal contacts. Thus, the ability to greatly reduce the recombination losses underneath the solar cell metal contacts (i.e., contact passivation) coupled with other technological advancements will be instrumental toward attaining the increasing solar cell efficiency targets.

One of the earliest examples of contact passivation can be found in the heterojunction silicon wafer solar cells, which utilizes a stack of intrinsic and doped amorphous silicon (a-Si:H) heterojunction layers [5–7] on both surfaces of the silicon wafer. The ultrathin (<5 nm) intrinsic a-Si:H layer not only serves to passivate the silicon surface but also to selectively enable hole or electron transport across this "tunnel layer," sandwiched between the overlying conductive a-Si:H layer and the crystalline silicon wafer. In this application, the contact-related recombination losses with the intrinsic/doped a-Si:H stack is significantly lower than utilizing the doped a-Si:H layers alone on the crystalline silicon wafers [5], hence establishing contact passivation for the former case. It can then be generalized that contact passivation can be established by deploying ultrathin passivating (and even in principal insulating, if thick) tunnel layers capped with a highly doped capping layer material with a suitable doping polarity or work function to form either hole-selective or electron-selective passivated contacts. Some examples of high/low work function capping layer materials such as transition metal oxides (WO_x, VO_x, etc.) and doped organic materials had been reported [4, 8].

Some prominent examples of single-junction silicon wafer-based high-efficiency (≥25%) solar cell concepts which adopt contact passivation include the amorphous silicon heterojunction interdigitated back contact (IBC) solar cell by Kaneka (26.6%) [9], the tunnel layer passivated interdigitated back contact (IBC) solar cell by SunPower (25.2%) [10], the polysilicon on oxide (POLO) passivated contact interdigitated back contact (IBC) solar cell by ISFH (26.1%) [11], and the conventionally front- and rear-contacted tunnel layer passivated contact solar cell (TOPCon) by the Fraunhofer ISE team (25.7%) [12]. The excellent performance of the TOPCon cell (despite being conventional front- and rear-contacted, instead of being contacted in an all-back-contact configuration) can be attributed to the highly effective and simplified full-area rear-side passivating contact scheme, which inserts an electron-selective tunnel layer passivated rear-side contact between the wafer and the full-area rear-side contact of the solar cell, comprising a wet-chemically formed silicon oxide tunnel layer (wet-SiO_x) and a highly n-doped polysilicon capping layer. This achieves both excellent interface passivation toward the silicon wafer and a highly selective collection of excess electron charge carriers. Although this work was established on a small-sized (4 cm^2) float-zone n-type silicon wafer, adopting a conventional front-side selective emitter, photolithography processes, and evaporated contacts, it has set the stage for immense research interests such as those reported in Refs. [12–26]. Contact passivation presents a clear advantage over the popular passivated emitter rear contact (PERC) solar cell concept by UNSW [27], which is currently a large scale adopted by the industry (as of Jan. 2019), as an even higher solar cell efficiency can be reached (i.e., by directly passivating the metal solar cell contacts instead of "only" reducing the metal contact area fraction).

An ideal tunnel layer, suited for contact passivation, (i) exhibits a tunneling relevant thickness (i.e., <2 nm) [14], (ii) exhibits excellent interface passivation toward the crystalline silicon wafer [28, 29], and (iii) contributes only minimally to the total contact resistance of the solar cell (in the order of maximal 1 Ω cm^2) [30]. Furthermore, an ideal capping layer, suited for contact passivation, should be either (i) highly doped or (ii) exhibit a high/low work function [31] in order to ensure selective excess charge carrier extraction.

The already proven success on electron-selective passivated contacts is also generating huge interest and research activities on hole-selective passivated contacts now. Pertaining to the feasibility studies of different tunnel layer candidates for hole-extracting passivated contacts, most previous reports had focused on using silicon-based oxides formed via either wet-chemical approaches (wet-SiO_x) or UV/ozone photo-oxidation (ozone-SiO_x) approaches. In our published works [28, 29, 32–34], a comprehensive evaluation of passivation quality and interface properties of silicon-based oxides (SiO_x) and atomic layer-deposited aluminum oxides (ALD-AlO_x) had revealed a larger potential for ALD-AlO_x to be integrated in hole-selective passivated contacts as compared to the commonly used wet-SiO_x or ozone-SiO_x. This stems from a significantly higher negative fixed interface charge density (~1 order of magnitude higher at -6×10^{12} cm^{-2}) even at a tunneling relevant thickness (just a few ALD cycles) while maintaining a relatively low interface defect density (D_{it}) of ~2×10^{12} cm^{-2} eV^{-1}, which is comparable to the D_{it} of SiO_x-based tunnel layers. The high negative fixed interface charges of the ALD-AlO_x tunnel layer will accumulate holes at the c-Si interface, which will simultaneously enhance hole extraction probability and reduce surface recombination rates due to an efficient field-effect passivation in addition to the chemical passivation at the interface, as evident from the higher measured effective carrier lifetime (two orders of magnitude higher) than the passivation by either wet-SiO_x or ozone-SiO_x alone on symmetrically tunnel layer passivated n-type Cz wafers in our previous work [28]. These findings were consistent with literature for much thicker AlO_x layers [35–39]. For hole-extracting capping layer materials, various candidates had been suggested, which includes highly p-doped polysilicon, transition metal oxide films with high work function such as molybdenum oxide (MoO_x) [40–45], tungsten oxide (WO_x), vanadium oxide (V_2O_5), cuprous oxide (Cu_2O) [46], or alternatively organic polymers, such as poly(3,4-ethylenedioxythiophene): poly(styrenesulfonate) (PEDOT:PSS) [47–49], among others. It is worthy to highlight that the transition metal oxide films exhibit a tunable work function between 4.7 and 7 eV [50, 51] by an appropriate combination of materials, while organic polymers can also exhibit a tunable work function from 3.0 to 5.8 eV by chemical modification [8]. As an example, Zielke et al. [52] has demonstrated a cell efficiency of 18.3 and 20.6% for both n-type silicon and p-type silicon solar cells, respectively, which deploys a rear-side tunnel layer passivated hole-extracting metal contact using their specifically adapted organic PEDOT:PSS blend as capping layer. Such findings could open up new opportunities for potentially low-cost novel material integration for high-efficiency solar cell concepts in the future.

Regarding contact passivation, however, it is to be noted that in most of these reports, the carrier-selective passivated contacts were mostly deployed at the rear side of the solar cell, while the front side composes of a conventionally diffused silicon surface followed by the standard anti-reflection coatings and screen-printed fire-through metal contacts. Since the rear-side deployed passivated contacts can achieve an excellently low contact recombination loss, instinctively the next focus will be to reduce the contact recombination loss at the front side as well in order to improve device performance. With varying degrees of success using either electron-selective or hole-selective passivated contacts in a standalone configuration, the question arises on the feasibility to integrate both electron-selective and hole-selective passivated contacts together in a typical silicon solar cell architecture. Regardless of the technological advances, the fundamental driving factors toward industry adoption will still be the same as outlined earlier (i.e., cost-effectiveness of the material and processes, manufacturing scalability, device performance, and product stability). Hence, it is of keen interest in this paper to evaluate the feasibility of combining our optimized electron-selective and hole-selective passivated contacts obtained via industrial relevant processes onto an otherwise conventional front and rear screen-printed silicon solar cells and comparing that to solar cells with only a rear-side passivated contact scheme.

In this work, we will investigate "conventional" SiO_x/poly-Si passivated contacts to be deployed on both sides of the solar cell, instead of only being deployed rear side. Using different lifetime test structures and solar cell structures, the following topics are investigated: (i) the influence of the tunnel oxide choice on the passivation quality, comparing wet-chemically formed oxides (wet-SiO_x), UV photo-oxidation-formed ozone oxides (ozone-SiO_x), and in situ thermal oxidation-formed oxides (thermal-SiO_x); (ii) the impact on the contact passivation quality after doped silicon capping layers were applied upon the tunnel oxide layers on the same lifetime test structures (formed via tube diffusion doping of low-pressure chemical vapor deposition (LPCVD) of intrinsic polysilicon layers, to serve as either electron-selective or hole-selective capping layers); (iii) the influence of the surface conditions on the passivation quality by both types of electron- or hole-selective passivated contacts; (iv) the integration of the optimal passivated contacts onto a practical double-sided passivated contact solar cell structure and studies on the resulting passivation quality, both prior to and after subsequent anti-reflection passivation layers (i.e., SiN_x layers) were applied; (v) the influence of the capping layer thickness on the absorbable cell current and various parasitic absorption losses via numerical analysis (SunSolve™) and our experimental approaches to realize ultrathin poly-Si capping layers; and (vi) the ability to apply screen-printed metal contacts on the developed electron-selective and hole-selective passivated contacts.

In addition, from the measured passivation quality results on lifetime test structures, a numerical calculation of the practical solar cell efficiency potential adopting both of our developed electron-selective and hole-selective passivated contacts was performed by utilizing the measured saturation current density J_0 and the measured contact resistance R_c from our investigated tunnel layer passivated contact test structures. This work demonstrates the feasibility and attractiveness of using industrial relevant processes to develop device quality tunnel oxide/doped polysilicon passivated contacts for effective contact passivation on both textured and planar silicon surfaces. A major highlight of this work is the demonstration of a practical solar cell efficiency potential approaching 24% on a large area (6-inch wafer), by deploying the in-house developed passivated contact layers on both sides of an otherwise conventionally processed silicon solar cells with industrial screen-printed contacts.

2. Experimental details

Firstly, the in-house development of device quality passivated contacts based on wet-SiO_x/poly-Si(doped), ozone-SiO_x/poly-Si(doped), or in situ thermal-SiO_x/poly-Si(doped) stack was established using simple planar symmetrical lifetime test structures as sketched in **Figure 1**. Such structures are convenient for assessing (i) the resulting tunnel layer/doped capping layer stack thickness; (ii) the passivation quality, attributing from the passivated contacts alone (i.e., determining minority carrier lifetime τ_{eff}, reverse saturation current density J_0, and implied open-circuit voltage iV_{OC}); and (iii) the tunneling resistance (i.e., determining the contact resistance R_c). Starting from bare diamond-wire cut Cz silicon wafers (NorSun, 190 μm thick, and

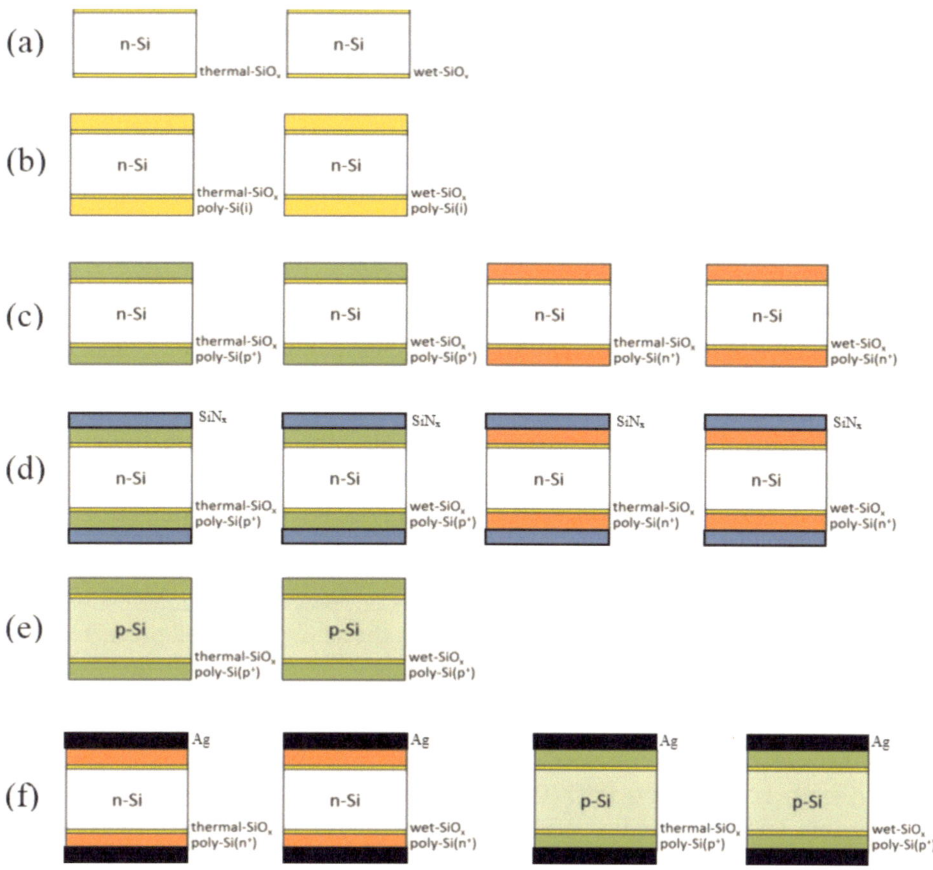

Figure 1.
Schematic of the (a-e) symmetrical lifetime and (f) contact resistance test structures utilized for assessing/optimizing the passivation quality and minimizing the contact resistance of both, electron-selective and hole-selective passivated contacts, developed on either a wet-SiO_x, ozone-SiO_x or an in-situ thermal-SiO_x tunnel layer. Using the sketched symmetrically passivated contact test structures, the developed SiO_x/poly-Si passivated contacts were characterized in terms of their passivation quality, doping profiles, film uniformity and contact resistance.

wafer resistivity of 3.4 Ω cm), these wafers received a saw damage etch removal process, followed by a standard RCA and HF clean process. The next step is the deposition of the various tunnel oxide layers (see **Figure 1(a)**). For lifetime test samples that require the wet-SiO_x tunnel layers, these samples were subjected to one more round of RCA2 process for 5 min (using deionized water, HCl, and H_2O_2 in the volume ratio of 0.84:0.08:0.08) in order to form the wet-SiO_x tunnel layer. Other selected lifetime test samples were deposited with a symmetrical ozone-SiO_x (UVO-Cleaner® 42, Jelight Company Inc.). The samples planned for an in situ thermal-SiO_x tunnel layer were processed using the low-pressure chemical vapor deposition (LPCVD) tool (TS-Series, Tempress) by flowing the oxidative gases prior to the deposition of the intrinsic poly-Si capping layers. Second, intrinsic poly-Si capping layers were deposited on top of all tunnel layers investigated, using LPCVD (TS-Series, Tempress) (see **Figure 1(b)**). These intrinsic poly-Si capped lifetime test structures were subsequently subjected to detailed doping optimization studies, using an industrial relevant high-throughput diffusion tool (Quantum, Tempress) to obtain device quality electron-selective and hole-selective passivated contacts (see **Figure 1(c)**). The increase in passivation quality after the deposition of an additional SiN_x passivation layer was studied using test samples as sketched in **Figure 1(d)**. In order to assess the total contact resistance R_c, some selected samples as sketched in **Figure 1(c)** and **(e)** (now using a p-type wafer instead of an n-type wafer) were further symmetrically contacted by thermally evaporated silver (System Control Technologies), i.e., processing symmetric Ag/poly-Si(n^+)/tunnel-oxide/n-Si-wafer/tunnel-oxide/poly-Si(n^+)/Ag samples to study electron extraction and Ag/poly-Si(p^+)/tunnel-oxide/p-Si-wafer/tunnel-oxide/poly-Si(p^+)/Ag samples to study hole extraction (see **Figure 1(f)**). In such samples, an ohmic straight-line dark I–V curve can be obtained, from which the contact resistance R_c on each side can be determined (after subtracting the resistance contribution from the silicon bulk).

Next, considering that typical silicon solar cells are either single-sided textured or symmetrically textured, it is relevant to explore the passivation quality when these developed passivated contacts are deployed on textured surfaces as well, while comparing that to planar references, as sketched in **Figure 2**. The objective is to identify the suitability of our developed electron-selective and hole-selective passivated contacts for textured surfaces and to determine the optimum configuration for a silicon solar cell considering contact passivation for both the front and rear surfaces.

It will be shown in later sections that the optimum double-sided passivated contact scheme can be realized by deploying the electron-selective passivated contacts (i.e., poly-Si(n^+)/tunnel oxide stacks) on the front textured surface while deploying the hole-selective passivated contacts (i.e., poly-Si(p^+)/tunnel oxide stacks) on the rear planar surface. Subsequently, the silicon solar cell precursors with the optimum double-sided passivated contact scheme were experimentally realized according to the process flow shown in **Figure 3** and characterized in terms of the passivation quality and doping profile, both prior to and after the standard anti-reflection/passivation dielectric coatings were deposited (i.e., step. 11 and 12, respectively) via microwave PECVD (MAiA, Meyer Burger), while comparing that to the symmetrical lifetime test structures. Selected samples were then subjected to

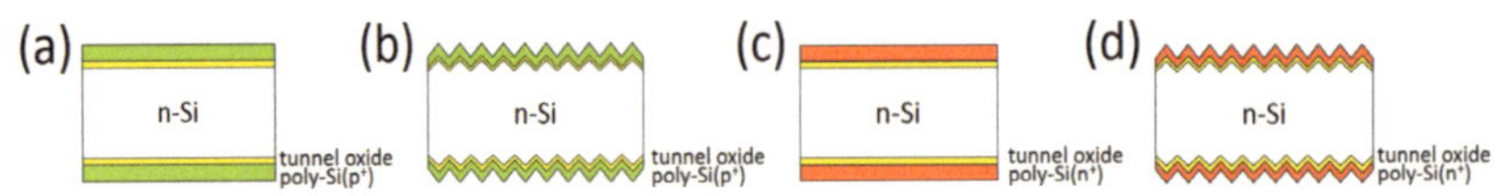

Figure 2.
Comparison of the passivation quality by both (a, b) hole-selective and (c, d) electron-selective passivated contacts on both planar and textured lifetime test structures.

DOI: http://dx.doi.org/10.5772/intechopen.85039

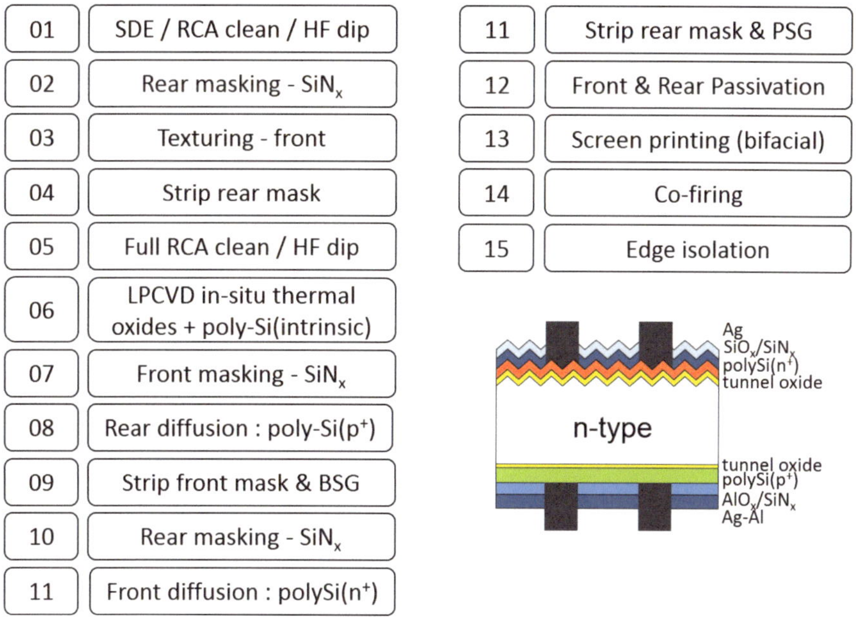

Figure 3.
Potential process flow for a silicon solar cell adopting double-sided passivated contacts and bifacial metal contacts.

a conventional full-area or bifacial screen printing process using commercially available fire-through paste to contact the electron-selective and hole-selective passivated contacts, through a high-temperature co-firing process at ~740°C in a fast-firing furnace (BTU) for 1 min. It is to be noted that the time of 1 min accounts for the total time spent within the fast-firing furnace, moving the intended sample across five temperature zones with increasing temperatures, with an estimated time of 5 seconds within the final peak temperature zone. As a final step, an edge isolation is carried out on the finished solar cell via a nanosecond laser process (ILS500LT, InnoLas), followed by electrical characterization.

One potential issue with replacing the conventional diffused regions with carrier-selective passivated contacts (such as the poly-Si(doped)/tunnel oxide stack in this work) is the presence of parasitic absorption, similar to the case of transparent conductive oxides or amorphous silicon layers in a heterojunction silicon wafer solar cell concept. Hence, there is an optimization potential toward simultaneously achieving excellent passivation quality of both textured and planar surfaces while minimizing the doped poly-Si capping layer thickness as much as possible in order to minimize the parasitic absorption issue.

Thus, it has been tested experimentally how thin our developed contact passivation layers can become while maintaining their excellent passivation quality. This has been realized by two different experimental approaches: (1) applying etch-back technology, thereby thinning down the already optimized thick layers, and (2) diffusion re-optimization for ultrathin LPCVD of intrinsic poly-Si layers.

To provide more insights into the influence of the doped capping layer thickness on the maximum absorbable current density $J_{\text{absorbed, cell}}$ attainable in a silicon solar cell, the simulation program SunSolve™, available on PV Lighthouse [53], was utilized. The SunSolve™ calculator combines Monte Carlo ray tracing with thin-film optics to calculate the maximum potential photogeneration current in the solar cell for the standard AM1.5G spectrum, as well as the corresponding optical losses occurring elsewhere (i.e., front-reflected, front-escaped, rear-escaped, parasitic absorption, edge absorption). Using SunSolve™, we provide a quantitative discussion of the influence of our doped poly-Si capping layer thickness on $J_{absorbed,\ cell}$ and $J_{absorbed,\ parasitic}$ which will ultimately affect the solar cell performance.

Finally, it is of keen interest to predict the impact of combining both of our developed electron-selective and hole-selective passivated contacts on the rear (and front side) of a silicon solar cell. To do this, we utilized Brendel's model [54] to predict the efficiency potential of a passivated contact and further enhanced the model to explicitly consider front-side conventional screen-printed contacts. This is done by additionally considering the combined front-side saturation current density $J_{0,\ front}$ (contributed by both the front-side metal-contacted regions and metal-passivated regions) and the front-side contact resistance $R_{c,\ front}$ of the screen-printed contacts. Thus, practical iso-efficiency contour plots (under a variation of the $J_{0,\ rear}$ and $R_{c,\ rear}$ values of the rear-side passivated contact) can be obtained, allowing us to predict a practical solar cell efficiency potential, given known $J_{0,\ front}$, $R_{c,\ front}$, $J_{0,\ rear}$, $R_{c,\ rear}$ values. Subsequently, our individually measured $J_{0,\ rear}$ and $R_{c,\ rear}$ values for our investigated passivated contacts in this work were inserted into this iso-efficiency contour plot, and a realistic prediction of the solar cell efficiency potential can be realized for both single-sided passivated contact scheme and double-sided passivated contact schemes.

Last but not least, the feasibility to contact our developed ultrathin contact passivation layers by an industrially suited method (i.e., aiming at conventional screen printing) is investigated, and the remaining issues, still to be solved in order to reach this goal, are addressed.

Concerning characterization metrology, we used the following tools: The average thickness and uniformity of the tunnel layers/doped poly-Si capping layers were determined by ellipsometry (SE-2000, Semilab) over a 9-point mapping measurement. The passivation quality was determined from the injection-dependent effective carrier lifetime measurements using a contactless flash-based photoconductance decay tester (WCT-120, Sinton Consulting) operated in both transient and quasi steady-state modes (QSSPC), which adopts an intrinsic carrier concentration of 8.6 $\times 10^9$ cm^{-3} in the calculation of the saturation current densities. To provide further insights at the tunnel layer/silicon interface, the fixed interface charge density Q_f and the interface defect density distribution $D_{it}(E)$ were determined using time-resolved contactless corona charge—Kelvin probe measurements (PV-2000, Semilab). Considering our ultrathin dielectrics, and its high potential for charge leakage, PV-2000 utilizes a "Self-Adjusting SteadyState Technique (SASS)" which takes into consideration the SASS voltage obtained using both positive and negative corona charges in order to calculate a leakage index (equivalently a correction factor) which accounts for the dielectric leakage when present and applicable to both ultrathin and thicker dielectrics, in order to produce a reliable representation of the Q_f and $D_{it}(E)$ values across different samples. The film structure of the doped silicon capping layer was determined via Raman spectroscopy (SE-2000, Semilab). The light and dark I–V measurements were performed on an LED-based AAA-calibrated I–V tester (Sinus220, Wavelabs).

3. Results and discussion

3.1 Screening of tunnel oxide layers for contact passivation

As mentioned earlier, the development and optimization of contact passivation layer stacks were initiated on symmetrical planar lifetime test structures as sketched in **Figure 1**. Prior to deposition of the doped capping layer, various tunnel oxide candidates were screened (i.e., wet-SiO_x, ozone-SiO_x, and thermal-SiO_x) in terms of their deposition techniques as well as the time required to get tunneling relevant thicknesses. Starting from our wet-chemically formed oxides (wet-SiO_x) via the standard RCA2 solution, **Figure 4(a)** shows that the resulting wet-SiO_x tunnel oxide thickness is independent of the oxidation time utilized (1–10 min) and is well within the tunneling relevant thickness regime (~1.2–1.5 nm). These wet-SiO_x tunnel oxide layers were also found to exhibit a highly leaky interface toward the silicon bulk, as attempts to determine the wet-SiO_x/c-Si interface properties Q_f and $D_{it}(E)$ were unsuccessful due to its inability to retain the deposited corona charges. Nonetheless, this is likely to be beneficial for the subsequent charge collection process, since the charge carriers to be collected can easily tunnel through such an oxide layer. It is also worthy to note that these symmetrical planar lifetime structures with only a wet-SiO_x tunnel oxide do not passivate well ($\tau_{eff} \sim 4$ μs) and are only effective when subsequently coupled with a highly doped capping layer to form a wet-SiO_x/poly-Si(doped) passivated contact scheme, as will be shown in the coming sections.

In contrast, for the investigated UV/ozone photo-oxidation-formed ozone-SiO_x tunnel oxides, **Figure 4(b)** shows that the resulting ozone-SiO_x layer thickness shows a time dependence of the photo-oxidation time, which increases from ~1.3 nm for an exposure time of 3 min to ~2.5 nm for 10 min. Beyond 10 min, the thickness of the ozone-SiO_x layer saturates at ~2.7 nm (i.e., surface reaction limited). Hence, considering the need for tunneling relevant applications (<1.5 nm), the UV/ozone exposure time should be limited to ≤3 min. Similar to the wet-SiO_x case, the ozone-SiO_x tunnel oxides were also found to be leaky in the as-deposited state, evident from its inability to measure Q_f and $D_{it}(E)$. In terms of passivation, symmetrically ozone-SiO_x passivated planar lifetime samples also do not passivate well ($\tau_{eff} \sim 2$ μs) and should be coupled with a highly doped capping layer to form an effective contact passivation scheme as well.

Finally, our investigated in situ thermal oxides were also found to exhibit a deposition time dependence on the measured oxide thickness, in which an in situ oxidation time of 30 secs at 570°C was already sufficient to achieve a tunneling relevant thickness of 1.0 to 1.2 nm. At higher deposition timings (e.g., 5 min), the thickness increases to ~13 nm which is not suitable for tunneling relevant applications.

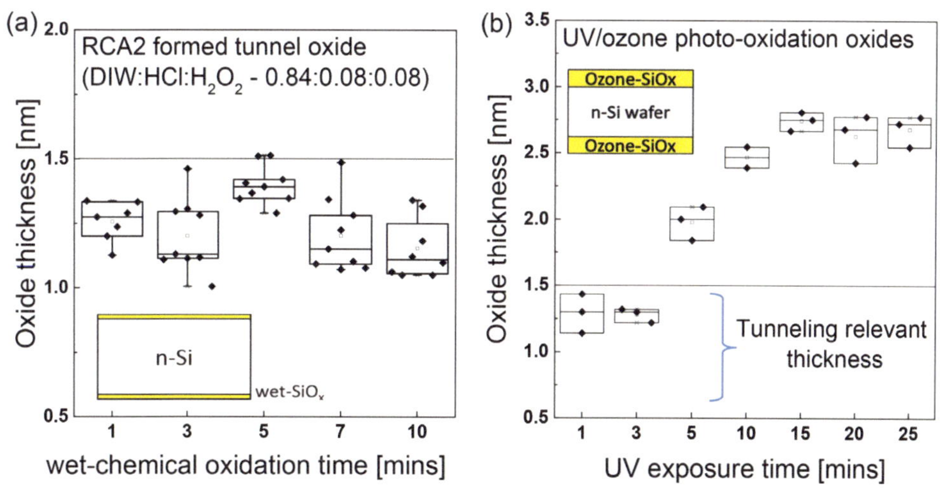

Figure 4.
(a) Comparison of the wet-chemical (RCA2) oxidation time on the measured wet-SiO_x tunnel oxide thickness. The wet-SiO_x thickness does not exhibit a time dependence (1–10 min) and has a thickness range of ~1.2–1.5 nm, relevant for device integration. (b) For ozone-SiO_x, the UV exposure time directly affects the ozone-SiO_x tunnel oxide thickness, with a recommended exposure time of 3 min to achieve tunneling relevant thickness.

Correspondingly, an in situ thermal oxide growth rate of ~2.4–2.6 nm/min can be expected. Interestingly, in contrast to the wet-SiO_x and ozone-SiO_x tunnel layers, our as-deposited thermal-SiO_x tunnel oxides were able to retain the deposited charges from the contactless corona charge—Kelvin probe measurements, allowing the fixed interface charge density and the interface defect density distribution to be determined (see **Figure 5**). At the first glance, this already suggests that the in situ thermal-SiO_x exhibits a higher film quality (i.e., non-leaky) than both wet-SiO_x and ozone-SiO_x. It is also likely that the thermal-SiO_x film structure is more dense, which can be beneficial when coupled with a highly doped silicon capping layer, which could reduce the out-diffusion of dopants into the c-Si bulk. **Table 1** summarizes the measured Q_f and D_{it} for our investigated tunnel oxides (wet-SiO_x, ozone-SiO_x, thermal-SiO_x) and compares that to literature-reported values. As plotted in **Figure 5**, and summarized in **Table 1**, the as-deposited in situ thermal-SiO_x (~1.2 nm) in this work exhibited a Q_f of -4.3×10^{11} cm^{-2} and a minimum D_{it} of ~2.5×10^{12} cm^{-2} eV^{-1}. The energy distribution of the interface defect density $D_{it}(E)$ as a function of the silicon band-gap energy for our in situ thermal-SiO_x layers is plotted in **Figure 5(b)**, showing a minimum D_{it} closer to the valence band, instead of the midgap position. This could be due to the increase of surface micro-roughness from the processing conditions, leading to a higher density of dangling bond defects in the higher part of the silicon energy

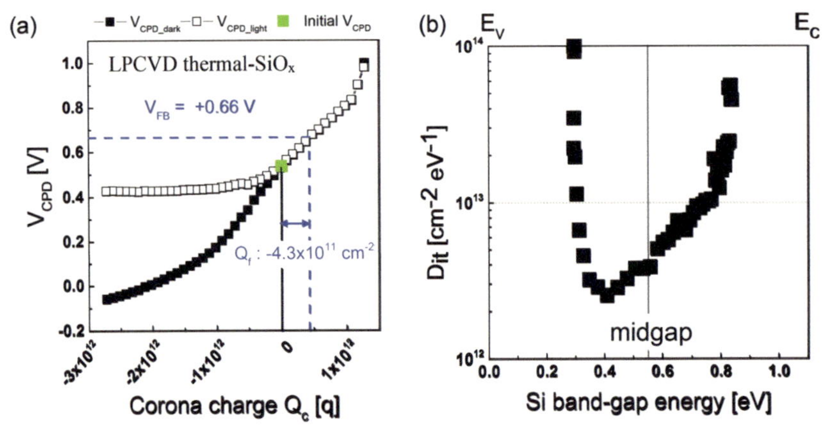

Figure 5.
The in-situ thermal silicon oxides deposited within the LPCVD process prior to the intrinsic poly-Si layers exhibited (a) a negative fixed interface charge density Qf of -4.3×10^{11} *cm*$^{-2}$ *and (b) a minimum interface defect density* $D_{it(min)}$ *of* ~2.5×10^{12} *cm*$^{-2}$ *eV*$^{-1}$.

Tunnel oxide	Thickness (nm)	Q_f (cm^{-2})	D_{it} (cm^{-2} eV^{-1})	References
Thermal-SiO_x	~1.2	-4.30×10^{11}	2.50×10^{12}	This work
Wet-SiO_x	~1.5	Not measurable	Not measurable	This work
Ozone-SiO_x	~1.3	Not measurable	Not measurable	This work
Thermal-SiO_x	~50–240	$+3.00 \times 10^{11}$	10^{10} ~ 7×10^{11}	[56–59]
Wet-SiO_x	~1–2	$+1.28 \times 10^{12}$	5.17×10^{12}	[57, 60–64]
Ozone-SiO_x	~1–2	No data	1.00×10^{13}	[55]
ALD-AlO_x	~1.5	-6.10×10^{12}	2.70×10^{12}	[28]

Table 1.
Comparison of the fixed interface charge density Q_f *and the interface defect density distribution* $D_{it}(E)$ *for different investigated tunnel oxides (thermal-*SiO_x*, wet-*SiO_x*, ozone-*SiO_x*) in this work as compared to literature.*

gap, similar to the observation by Angermann et al. [55] who observed a skewing toward the conduction band when p-type Si substrates are utilized.

As compared to other thermally grown silicon oxides [56–59] which exhibited significantly lower D_{it} values (~1–2 orders), their film thickness was however also significantly higher at ~50–240 nm, making it inappropriate for tunnel layer applications. On the other hand, the wet-SiO_x and ozone-SiO_x tunnel oxides reported in Refs. [55, 57, 60–64] do exhibit measurable Q_f and D_{it} values, unlike our investigated samples, which can be attributed to the deposition method and the post-deposition annealing conditions. Our wet-SiO_x and ozone-SiO_x tunnel oxides were unable to retain the deposited corona charges due to its leaky interface, which nonetheless could be beneficial for the purpose of tunneling carrier transport. Another noteworthy tunnel oxide candidate is atomic layer deposition (ALD) of aluminum oxide (AlO_x), whereby in one of our earlier publications [28], we experimentally realized ultrathin ALD-AlO_x films in the tunneling regime (~1.5 nm) which is capable of exhibiting a significantly higher negative Q_f of -6×10^{12} cm^{-2} and a D_{it} of ~2.7×10^{12} cm^{-2} eV^{-1}. This resulted in a ~110-fold increase in the initial passivation quality prior to the doped capping layers as compared to a conventional wet-SiO_x tunnel oxide layer using our test structures (i.e., from 2 to 218 μs) [28]. This finding positions ultrathin ALD-AlO_x as a highly attractive tunnel oxide candidate for hole-extracting selective contacts. In contrast, although the D_{it} values of our in situ thermal-SiO_x and ALD-AlO_x films are comparable, the thermal-SiO_x in this work exhibited one order lower negative fixed interface charge density which do suggest a reduced field-effect passivation and overall surface passivation prior to the doped capping layers, which is observed experimentally as well ($\tau_{eff} \sim 4.7$ μs). Nevertheless, this positions thermal-SiO_x as a tunnel oxide candidate suitable for both electron-extracting and hole-extracting selective contacts.

3.2 Screening LPCVD poly-Si capping layers for contact passivation

As compared to the TOPCon approach by the Fraunhofer ISE's team, which deposited doped amorphous silicon films followed by a suitable annealing condition and hydrogenation process to convert the highly doped amorphous silicon to highly doped polysilicon capping layers, we implement an alternative approach by first depositing intrinsic silicon films via the LPCVD approach, followed by either a phosphorus or boron diffusion process to convert it to a highly doped poly-Si(n^+) or poly-Si(p^+) capping layer, respectively. The optimization goal is to incorporate as much active dopants within the poly-Si layers as possible while reducing or avoiding the out-diffusion of dopants into the c-Si wafer bulk, which will increase the surface recombination rates and reduce the device performance, as also reported in Ref. [65].

As a start, Raman spectroscopy was utilized to monitor the structural evolution of our in-house deposited silicon capping layers, both in the as-deposited intrinsic case and after the optimal diffusion process (boron or phosphorus doped). **Figure 6** shows that our LPCVD as-deposited intrinsic silicon films were amorphous in film structure, evident by a single Raman peak centered at a Raman shift of ~480cm^{-1} [66]. Nonetheless, upon either a boron diffusion process or a phosphorus diffusion process, which takes place at temperatures between 850 and 950°C, these doped silicon films fully crystallize as evident by a single Raman peak centered at a Raman shift of ~520.5 cm^{-1} with a full width at half maximum (FWHM) of 5.3 and 4.0 cm^{-1}, respectively. These findings were comparable to our crystalline silicon wafer reference (Raman shift centered at ~520.6 cm^{-1} and a FWHM of 3.5 cm^{-1}). The slightly higher FWHM measured for our doped silicon films indicated a marginally higher structural disorder than a perfect crystalline silicon wafer bulk which

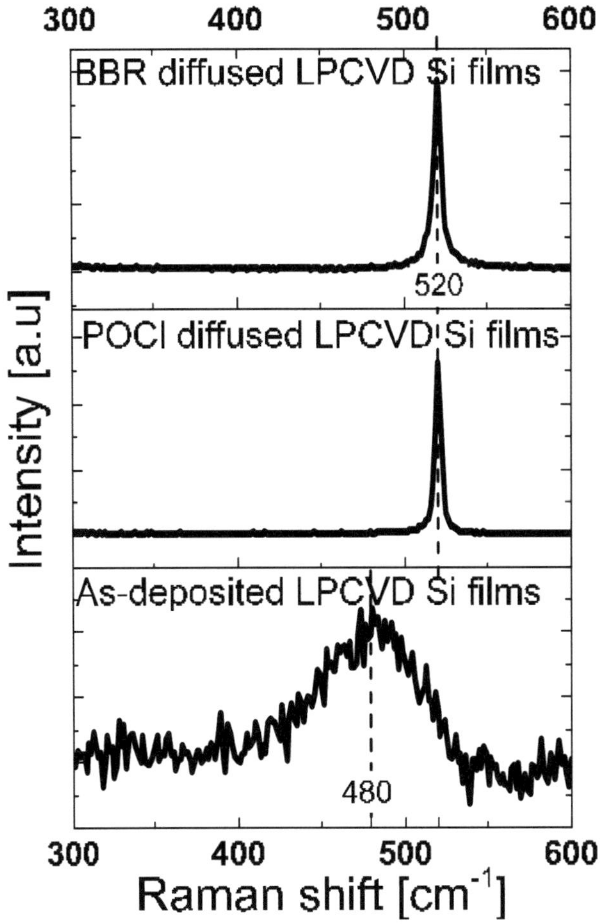

Figure 6.
Raman spectra for our in-house LPCVD of silicon films (~250 nm), comparing the film crystallinity in the as-deposited state, post-$POCl_3$ diffusion process, and post-BBR_3 diffusion process.

is not too surprising, given the high quantities of dopants (10^{19}–10^{20} cm^{-3}) incorporated in the former.

The corresponding dopant profile within these highly doped silicon capping layers can be extracted from ECV measurements as shown in **Figure 7**. After an optimized diffusion doping process to convert the thermal-SiO_x/a-Si(intrinsic) capping layer stack toward either an electron-selective passivated contact (i.e., thermal-SiO_x/poly-Si(n^+) stack) or a hole-selective passivated contact (i.e., thermal-SiO_x/poly-Si(p^+) stack), the ECV measurements revealed a peak doping concentration within the poly-Si(n^+) and poly-Si(p^+) capping layers as $\sim 1.5 \times 10^{20}$ and $\sim 5 \times 10^{19}$ cm^{-3}, respectively.

The tunnel oxide in the passivated contact stack not only serves as passivation/tunneling purposes, but it also likely serves as a blocking layer to reduce the out-diffusion of dopants from the highly doped silicon capping layer into the crystalline silicon wafer bulk. The lower active dopant concentration within the poly-Si(p^+) layer can be partially attributed to the lower doping efficiency of boron atoms than phosphorus atoms [67] based on the theoretical prediction of impurity formation energies and partially attributed to the higher diffusivity of the boron dopants [68] into the silicon bulk which resulted in a deeper boron-diffused junction (see **Figure 7**). Similar to other reports [65], we also observed experimentally that it is preferable to concentrate all the dopants within the poly-Si layers, as the

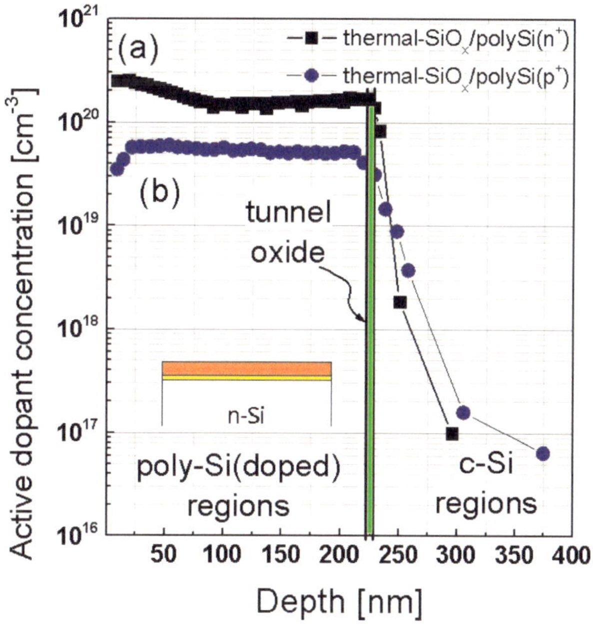

Figure 7.
*ECV profiles for both (a) electron-selective passivated contacts comprising thermal-SiO_x/poly-Si(n^+) stacks and (b) hole-selective passivated contacts comprising thermal-SiO_x/poly-Si(p^+) stacks. The electron-selective passivated contacts exhibited a higher peak doping concentration than the hole-selective counterpart by a factor of ~2 times. Poly-Si(p^+) layers also exhibited a higher out-diffusion of dopants into the c-Si bulk than the poly-Si(n^+) layers, which was found to limit the potentially achievable implied-V_{OC} values (see **Table 3**).*

out-diffusion of dopants is expected to lead to increased surface recombination rates and a corresponding drop in the overall passivation quality as well.

Table 2 summarizes our measured passivation quality results on planar symmetrical lifetime test structures with the optimized doped poly-Si(n^+) capping layers on various investigated tunnel oxide candidates (i.e., wet-SiO_x, ozone-SiO_x, thermal-SiO_x). **Table 2** shows that using planar symmetrical lifetime test structures, our wet-SiO_x/poly-Si(n^+) and ozone-SiO_x/poly-Si(n^+) passivated samples were exhibiting implied-V_{OC} values of ~719 mV and J_0 of ~6–9 fA cm^{-2}, while the thermal-SiO_x/poly-Si(n^+) passivated contact stack exhibited an even higher implied-V_{OC} values of 729 mV, despite a similar J_0 of ~9 fA cm^{-2}. The enhanced implied-V_{OC} values for thermal-SiO_x/poly-Si(n^+) passivation stack were consistent with the earlier discussion on the tunnel oxides, in which a thermal-SiO_x tunnel oxide is likely more effective in reducing the out-diffusion of phosphorus dopants from the poly-Si(n^+) into the c-Si wafer bulk, hence providing better overall passivation quality (implied-V_{OC} increases by ~10 mV).

Since a typical silicon solar cell would be further coated with suitable anti-reflection layers (such as SiN_x or AlO_x/SiN_x stacks) prior to metallization, the influence of these layers on our symmetrical lifetime samples were evaluated as well, by capping the passivated contacts with an additional ~70-nm-thick SiN_x films symmetrically and its resulting passivation quality evaluated. As summarized in **Table 2**, the measured passivation quality further improves with the additional SiN_x capping layers upon all three investigated lifetime test structures with electron-selective passivated contacts. In particular, the thermal-SiO_x/poly-Si(n^+)/SiN_x-capped lifetime structure exhibits high implied-V_{OC} approaching 740 mV, with single-sided J_0 values down to ~2.5 fA cm^{-2}, which is already on par with the best results from the Fraunhofer ISE team [69]. Concurrently, similar studies were conducted on lifetime test structures with hole-selective passivated contacts, and selected results are highlighted in **Table 3**, which demonstrates the potential of our developed hole-selective contact passivation layers as well (i.e., thermal-SiO_x/poly-Si(p^+) or ALD-AlO_x/poly-Si(p^+) stacks) with implied-V_{OC} approaching 700 mV in the as-deposited state and a further enhancement to 713 mV with single-sided J_0 values down to ~4 fA /cm^{-2} after applying symmetrical SiN_x capping layers. This can be attributed to the hydrogenation process which occurs spontaneously during the deposition of the SiN_x capping layer, which helps to reduce the interface defect densities and directly improves the passivation quality [70]. Comparing our results to the excellent results from the Fraunhofer ISE team [69], which adopts PECVD of p-doped a-Si:H layers followed by sintering and SiN_x capping (with a high implied-V_{OC} values up to 732 mV and single-sided J_0 values <1 fA cm^{-2}), we do identify optimization potential for our LPCVD of intrinsic silicon capping layer and the associated boron diffusion optimization thereafter.

Tunnel layer/capping layer	Method	iV_{OC} (mV)	Total J_0 (fA cm^{-2})	References
Wet-SiO_x/poly-Si(n^+)	PECVD Centrotherm	719 ± 2	–	FhG-ISE
Wet-SiO_x/poly-Si(n^+)	PECVD RF-MAiA	740	–	[69]
Wet-SiO_x/poly-Si(n^+)	LPCVD Tempress	719	~9	This work
Ozone-SiO_x/poly-Si(n^+)	LPCVD Tempress	719	~6	This work
Thermal-SiO_x/poly-Si(n^+)	LPCVD Tempress	729	~9	This work
After a hydrogenation/anti-reflection coating step by SiN_x				
Wet-SiO_x/poly-Si(n^+)/SiN_x	LPCVD + MAiA	730	~5	This work
Ozone-SiO_x/poly-Si(n^+)/SiN_x	LPCVD + MAiA	732	~4	This work
Thermal-SiO_x/poly-Si(n^+)/ SiN_x	LPCVD + MAiA	737	~5	This work
The thickness of the tunnel oxides/doped poly-Si layer/SiN_x layer is ~1.5/250/80 nm, respectively.				

Table 2.
Comparison of the passivation quality of electron-selective passivated contacts on planar Cz n-Si symmetrical lifetime samples, both prior to and after the additional hydrogenation process step via the symmetrical addition of the SiN_x capping layers.

Tunnel layer/capping layer	Method	iV_{OC} (mV)	Total J_0 (fA cm^{-2})	References
Thermal-SiO_x/poly-Si(p^+)	LPCVD Tempress	698	37	This work
ALD-AlO_x/poly-Si(p^+)	ALD Solaytec + LPCVD	697	~26	This work
After a hydrogenation/anti-reflection coating step by SiN_x				
Thermal-SiO_x/poly-Si(p^+)/ SiN_x	LPCVD +MAiA	713	~8	This work
Thermal-SiO_x/poly-Si(p^+)/ SiN_x	PECVD Centrotherm	732	~1	[69]
The thickness of the tunnel oxides/doped poly-Si layer/SiN_x layer is ~1.5/250/80 nm, respectively.				

Table 3.
Comparison of the passivation quality of hole-selective passivated contacts on planar Cz n-Si symmetrical lifetime samples, both prior to and after the additional hydrogenation process step via the symmetrical addition of the SiN_x capping layers.

3.3 Evaluation of developed contact passivation stacks on textured surfaces

Given the excellent passivation quality from our developed electron-selective and hole-selective passivated contacts on planar Cz silicon wafers, it is then of research and commercial interest to evaluate the performance of these layers on textured surfaces as well, to determine its viability for deployment on a conventional silicon solar cell structure which adopts a front-side textured surface and either a rear-side planar or textured surface. To evaluate that, the lifetime test structures as shown in **Figure 2** are utilized, featuring either symmetrical planar surfaces or symmetrical textured surfaces and symmetrically capped by either the electron-selective (thermal-SiO_x/poly-Si(n^+)) or hole-selective (thermal-SiO_x/poly-Si(p^+)) passivated contacts. The objective is to identify the suitability of our developed electron-selective and hole-selective passivated contacts for textured surfaces as well and to determine the optimum configuration for a silicon solar cell considering contact passivation for both the front and rear surfaces.

The highlight of this evaluation is plotted in **Figure 8**. Firstly, considering the influence of surface conditions on the passivation quality, it can be observed consistently from **Figure 8** and summarized in **Table 4** that both the electron-selective and hole-selective passivated contact stacks exhibited significantly better passivation quality on planar surfaces than on textured surfaces and which were consistent with the best results shown in **Tables 2** and **3**. Based on a batch average of 18 samples for each investigated lifetime test structure shown in **Figure 8**, the hole-selective passivated contacts on symmetrical planar lifetime test structures demonstrated an effective minority carrier lifetime τ_{eff} of ~1650 μs, a single-sided $J_{0,\ rear}$ of 27.5 fA cm^{-2}, and an implied-V_{OC} of 689 mV, which is a significant improvement over the textured case (τ_{eff} of ~170 μs, single-sided $J_{0,\ rear}$ of 265 fA cm^{-2}, and implied-V_{OC} of 628 mV). Effectively, upon deploying the hole-selective passivated contact on a textured surface, the τ_{eff} and implied-V_{OC} reduce by ~90 and ~8.9%, respectively. Similarly, while the electron-selective passivated contacts continued to exhibit excellent passivation quality on planar surfaces (τ_{eff} of ~6030 μs, single-sided $J_{0,\ rear}$ of 5.4 fA cm^{-2}, implied-V_{OC} of 723 mV), the passivation quality

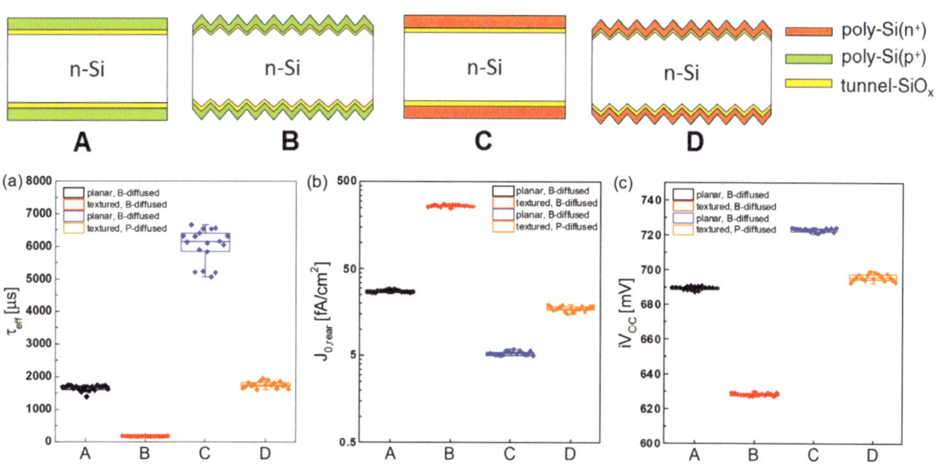

Figure 8.
*Comparison of the passivation quality (i.e. (a) effective carrier lifetime at 10^{15} cm^{-3} injection level, (b) rear side J_o values and (c) implied-V_{OC} values) by electron-selective (thermal-SiO_x/poly-Si(n^+)) and hole-selective (thermal-SiO_x/poly-Si(p^+)) passivated contacts on both symmetrical planar and symmetrical textured lifetime test structures. It can be observed that electron-selective passivated contacts are suitable for applications on both planar and textured surfaces (with implied-V_{OC} > 720 mV and > 695 mV respectively), while the hole-selective passivated contacts are only suitable for planar surfaces at the moment (with implied-V_{OC} approaching 700 mV, compare **Table 3**).*

Structure	Surface	Pass. contact type	τ_{eff} (μs)	$J_{0,\ rear}$ (fA cm^{-2})	iV_{OC} (mV)
A	Planar	Hole-selective	1649	27.5	689
B	Textured	Hole-selective	170	265	628
C	Planar	Electron-selective	6030	5.4	723
D	Textured	Electron-selective	1756	17.4	696

Table 4.
Summary of the average measured passivation quality for both electron-selective and hole-selective passivated contacts deployed on both symmetrical planar and symmetrical textured silicon lifetime test structures.

reduces on textured surfaces as well (τ_{eff} of ~1750 μs, single-sided $J_{0,\ rear}$ of ~17 fA cm^{-2}, implied-V_{OC} of 696 mV). Effectively, upon deploying the electron-selective passivated contact on a textured surface, the τ_{eff} and implied-V_{OC} reduce by ~71 and ~3.7%, respectively. Utilizing the same textured lifetime test structures, the electron-selective passivated contacts experience lower degradation of the passivation quality than the hole-selective passivated contacts by a factor of 2.4 times in terms of the implied-V_{OC} values. The lower passivation quality measured on textured surfaces is not too surprising, given that similar observations were observed when evaluating silicon dioxide thin-film passivation on either planar or textured surfaces [71]. In particular, this reduced passivation quality can be attributed to (i) increased surface area (~73% more surface area for textured [111] surfaces than planar [100] surfaces), (ii) increased density of dangling bonds at a [111] surface, and (iii) a higher concentration of interface defects, which could originate from the mechanical stress in the dielectric-silicon interfaces at creases, edges, or vertices [72]. Despite this inherent limitation, we demonstrate in this work that our electron-selective (thermal-SiO_x/poly-Si(n^+)) passivated contacts have a great potential for being deployed on both planar and textured surfaces, while our hole-selective (thermal-SiO_x/poly-Si(p^+)) passivated contacts are currently only suited on planar surfaces, based on our current developments.

Thus, if a double-sided contact passivation scheme is to be considered, the results in this work suggest that it is preferable to implement a solar cell structure with a textured front surface and a planar rear surface, and adopting the electron-selective passivated contacts at the textured front surface and the hole-selective passivated contacts at the planar rear surface, as will be shown in the next section.

3.4 Deployment of double-sided passivated contacts at the solar cell level

Based on the findings from the previous section, the deployment of double-sided passivated contacts at the solar cell level had been experimentally realized on n-type silicon wafers with a textured front surface and a planar rear surface and adopting an electron-selective (thermal-SiO_x/poly-Si(n^+)) passivated contacts at the textured front surface and a hole-selective (thermal-SiO_x/poly-Si(p^+)) passivated contacts at the planar rear surface. This is further compared to reference lifetime test structures with either symmetrical planar surfaces with symmetrical hole-selective passivated contacts or symmetrical textured surfaces with symmetrical electron-selective passivated contacts, as sketched in **Figure 9**. As shown in **Figure 9** and summarized in **Table 5**, the lifetime test structures within this second batch of samples processed similarly to **Figure 8** were able to consistently deliver excellent passivation qualities for the planar and textured lifetime test structures. In particular, **Table 5** shows that structure B (symmetrically planar lifetime test structures with symmetrical hole-selective passivated contacts) was able to again demonstrate an implied-V_{OC} of ~696 mV and a single-sided $J_{0,\ rear}$ value of ~19.5 fA cm^{-2}, while structure C

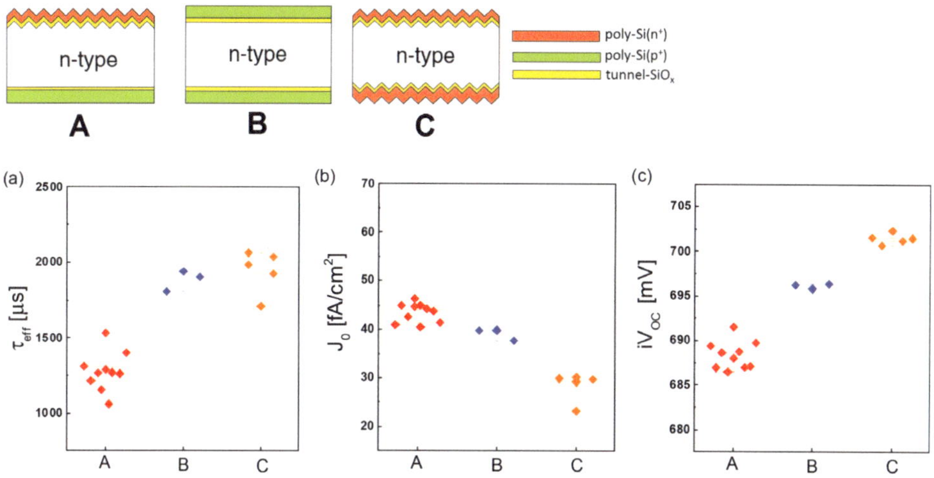

Figure 9.
Comparison of the passivation quality (i.e. (a) effective carrier lifetime at 10^{15} cm^{-3} injection level, (b) rear side J_o values and (c) implied-V_{OC} values) when both the electron-selective (thermal-SiO_x /poly-Si(n^+)) passivated contacts and hole-selective passivated contacts (thermal-SiO_x /poly-Si(p^+)) are deployed on both solar cell structure A (front-side textured, rear-side planar silicon wafer), lifetime test structure B (symmetrically planar), and lifetime test structure C (symmetrically textured).

Structure	Surface (poly-Si thickness)	Pass. contact type	τ_{eff} (μs)	Total J_0 (fA cm^{-2})	iV_{OC} (mV)
A	Front-textured (250 nm) Rear-planar (250 nm)	Front electron-selective Rear hole-selective	1273	43.4	688
B	Sym.-planar (250 nm)	Hole-selective	1883	39.1	696
C	Sym.-textured (250 nm)	Electron-selective	1943	28.5	701

Table 5.
Summary of the average measured passivation quality for both electron-selective and hole-selective passivated contacts deployed on different wafer surfaces.

(symmetrically textured lifetime test structures with symmetrical electron-selective passivated contacts) was able to demonstrate an implied-V_{OC} of ~701 mV and a single-sided $J_{0,\ rear}$ value of ~14 fA cm^{-2}. At the first thoughts, we would expect structure A (the double-sided passivated contact solar cell precursors) to exhibit a measured passivation quality that lies between that exhibited by structure B and structure C. However, the actual measured results revealed that structure A exhibited a poorer passivation quality than both structure B and structure C. None-theless, structure A was able to demonstrate quite high implied-V_{OC} of ~688 mV and a total J_0 value of ~43 fA cm^{-2}, prior to any anti-reflection/passivation layers, which likely cannot be attained by conventional diffusion of silicon solar cell precursors.

With a closer look at the key process steps, the key difference between the symmetrical lifetime test structures and the solar cell structures is that the former structures can be done in a one-step diffusion process, while the latter structures would require a series of dielectric masking to achieve single-sided diffused poly-Si layers with different polarities, starting from the higher-temperature requirement first (i.e., boron diffusion toward poly-Si(p^+) in this work), followed by the diffusion process with a lower-temperature requirement (i.e., phosphorus diffusion toward poly-Si(n^+)). The goal is to reduce the drive-in/out-diffusion of boron dopants from the poly-Si(p^+) layer into the silicon bulk which is expected to lead to an increased near-surface recombination and poorer passivation quality, as evident from our measurements as well (see **Figure 9**). **Figure 10** shows a comparison of the ECV profiles done on the same poly-Si(p^+) layer in the as-diffused state and after an additional diffusion masking and front-side phosphorus diffusion step. It can be clearly seen in the latter that the boron dopants have out-diffused from the poly-Si(p^+) capping layer into the silicon bulk, which is consistent with the reduced passivation quality measured on the solar cell precursors. Unfortunately, this issue is inevitable for our current investigated approach of obtaining the doped silicon capping layers, although the dopant out-diffusion could be better controlled via diffusion recipe optimization.

For a conventional silicon wafer solar cell, suitable dielectric thin films or stacks of thin films (such as SiO_x, SiN_x, AlO_x) would be deposited on the silicon wafer surfaces to serve as anti-reflection/passivation prior to the metallization step. Similarly, in this work, the double-sided passivated contact solar cell precursors shown in **Figure 9** were symmetrically capped with PECVD of ~70-nm-thick SiN_x films. The resulting passivation quality before and after additional SiN_x capping is plotted in **Figure 11** and listed in **Table 6**.

It can be seen from **Figure 11** that upon the deposition of an additional symmetrical SiN_x capping layer, there is a striking improvement in the pre-metallized solar cell precursors, in which the $\tau_{eff}/J_0/iV_{OC}$ values improve from 1.5 ms/48 fA cm^{-2}/ 690 mV to 2.4 ms/16.5 fA cm^{-2}/713 mV. This improvement can be attributed to the hydrogenation effects from the overlying SiN_x films, which is expected to further reduce the interface defect densities and improve its corresponding interface passivation quality, as evident from the measured lifetime results presented earlier. This observation was also consistently observed on the symmetrical lifetime test structures, in which the textured samples with electron-selective passivated contacts exhibited improvement in the $\tau_{eff}/J_0/iV_{OC}$ values from ~1.9 ms/28.5 fA cm^{-2}/701 mV to ~5.5 ms/13.3 fA cm^{-2}/731 mV, while the planar samples with hole-selective passivated contacts exhibited improvement in the $\tau_{eff}/J_0/iV_{OC}$ values from ~1.9 ms/39 fA cm^{-2}/696 mV to ~3 ms/30.8 fA cm^{-2}/710 mV.

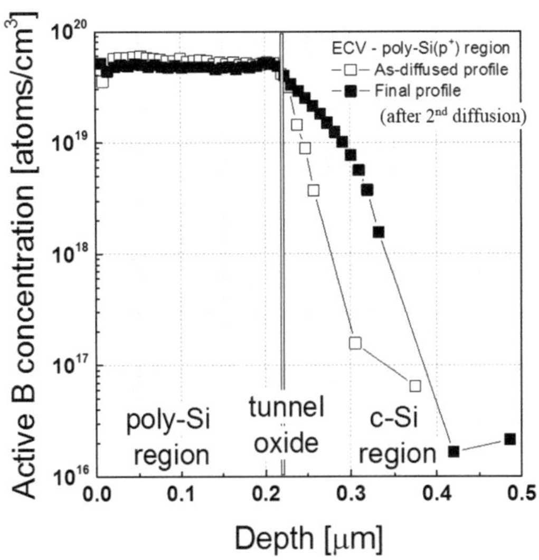

Figure 10.
Measured ECV profile for the poly-Si(p^+) region, comparing the as-diffused profile after the first rear-side boron diffusion (i.e., same compared to the lifetime test structure) and the final boron diffusion profile (i.e., after additional steps of masking, the second front-side phosphorus diffusion, and the chemical mask removal process). For the solar cell precursors, the additional high-temperature process step (second diffusion) causes out-diffusion of boron dopants from the poly-Si(p^+) layer into the silicon wafer bulk, as evident from ECV measurements.

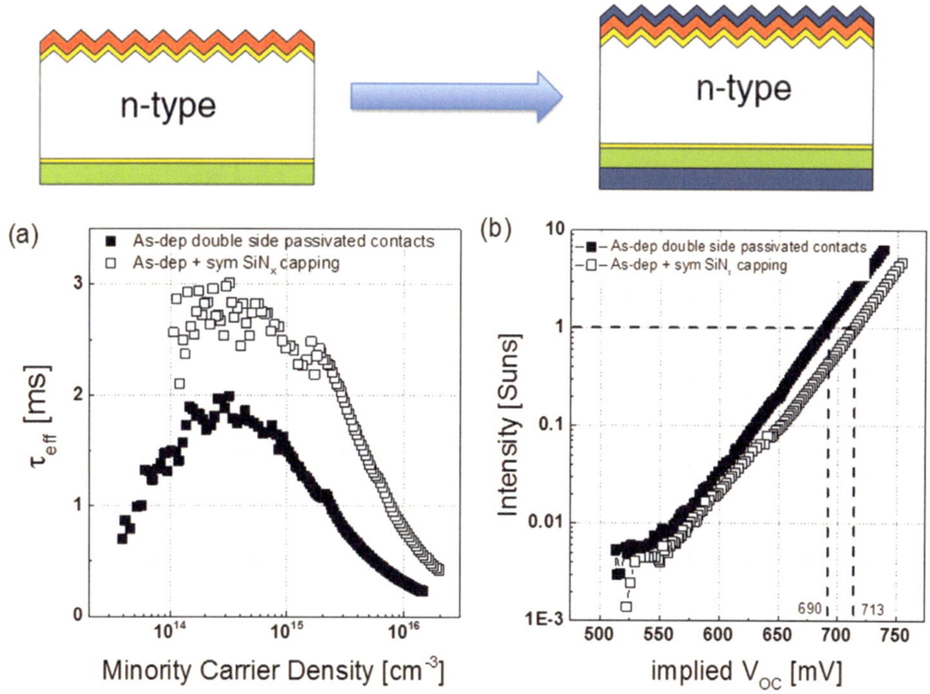

Figure 11.
Measured passivation quality of the deployed double-sided passivated contacts on the solar cell structure sketched in ***Figure 10****, both in the as-deposited state and after a symmetrical SiN_x capping layer, was applied. With the symmetrical SiN_x capping, which leads to the pre-metallized solar cell precursors, an excellent implied-V_{OC} of 713 mV was obtained. The corresponding total J_o values improved from 48 to 16.5 fA cm^{-2}, an approximately threefold improvement.*

Structure	Surface (poly-Si thickness)	Pass. contact type	τ_{eff} (μs)	Total J_o (fA cm^{-2})	iV_{OC} (mV)
Before SiN_x	Front-textured (250 nm) Rear-planar (250 nm)	Front electron-selective Rear hole-selective	1500	48	690
After SiN_x	Front-textured (250 nm) Rear-planar (250 nm)	Front electron-selective Rear hole-selective	2400	16.5	713

Table 6.
Summary of the measured passivation quality of a double-sided passivated contact solar cell precursor, before and after additional SiN_x capping.

3.5 Addressing the parasitic absorption issue for highly doped poly-Si layers

Despite the excellent passivation qualities from the developed passivated contacts, one of the key challenges identified for device integration is the issue of parasitic absorption by these highly doped poly-Si capping layers. This issue is found to be more critical when the layers are deployed at the front surface than the rear surface, as simulation studies will show in the later sections. Hence, in order to address the parasitic absorption issue, a thinning of the doped poly-Si thickness is necessary.

Two different experimental approaches have been investigated: (1) applying a slow silicon etch-back technology, thereby thinning down our already well-optimized thick layers, and (2) performing a diffusion re-optimization for ultrathin LPCVD of intrinsic poly-Si layers. The goal is to determine the threshold (lowest thickness) of the poly-Si films necessary to achieve the same excellent passivation quality as the thicker counterparts while reducing the parasitic absorption issue as much as possible.

Using our slow silicon etch (SSE) solution (DIW:KOH (3.5%):NaOCL (63.25%) at 80°C), an etch rate of ~0.1 nm/s was determined, which was consistently observed for both poly-Si(n^+) and poly-Si(p^+) capping layers. **Figure 12** highlights the influence of the resulting doped poly-Si capping layer thickness on the measured passivation quality.

Interestingly, for hole-selective passivated contacts, the passivation quality can be preserved for a poly-Si(p^+) thickness from a thick ~250 nm down to ultrathin layers of approximately ~3 nm, with measured τ_{eff} of ~1.5 ms and implied-V_{OC} of ~690 mV, respectively. This suggests that a simple SSE etch could be an effective approach to reduce the poly-Si(p^+) capping layer thickness to an ultrathin (i.e., some nm ony) level. However, for electron-selective passivated contacts, the passivation quality was preserved only from ~250 nm down to ~70 nm, with measured $\tau_{eff} > 6$ ms and implied-$V_{OC} > 720$ mV, respectively. A further thickness reduction (<70 nm) leads to a severe degradation of passivation quality. As an example, upon reduction of the poly-Si(n^+) layer from 69 nm to 47 nm, the measured τ_{eff} and implied-V_{OC} reduce by 86 and 6.5%, respectively. Hence, considering the preference to deploy electron-selective passivated contacts (thermal-SiO_x/poly-Si(n^+)) on the textured surface, we have to investigate alternative approaches (as outlined in the following) to obtain ultrathin poly-Si(n^+) capping layers suitable for device integration at the front textured surface of a double-sided passivated contact solar cell.

One of the alternative approaches to obtain ultrathin poly-Si(n^+) layers is to directly deposit an ultrathin intrinsic poly-Si capping layer, followed by a further optimization of the phosphorus diffusion conditions. The goal is to obtain a highly doped thin poly-Si(n^+) capping layer which can achieve excellent passivation quality similar to the thicker poly-Si(n^+) counterparts while minimizing the in-diffusion of phosphorus dopants into the silicon bulk. To achieve this, ultrathin (~10 nm)

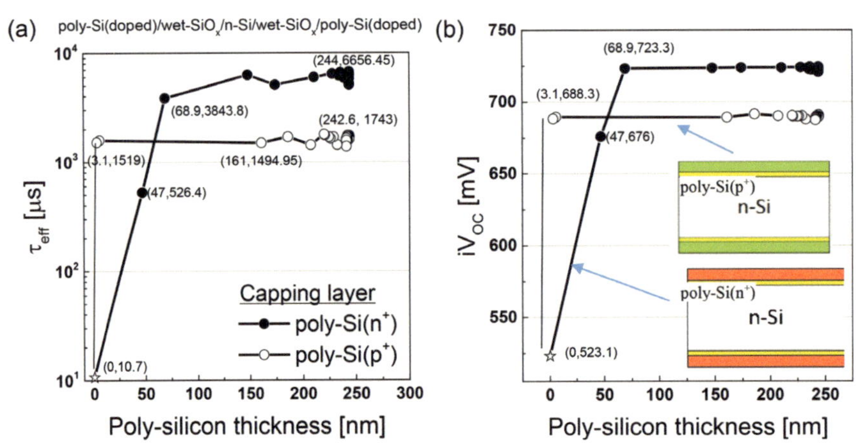

Figure 12.
Influence of the decreasing doped poly-Si capping layer thickness via the slow silicon etch process on the measured (a) minority carrier lifetime τ_{eff} and (b) implied-V_{OC} values for symmetrically planar lifetime test structures. Promising results are observed on hole-selective passivated contacts, in which the passivation quality is preserved for a poly-Si(p^+) capping layer thickness reduction from a thick ~250 nm down to a thin ~3 nm. In contrast, the passivation quality of the electron-selective passivated contacts with poly-Si(n^+) capping layer was preserved down to a thickness of ~70 nm, beyond which there is a drastic drop in passivation quality. The "star" symbol refers to the case where there is no doped poly-Si capping layer (i.e., only the tunnel oxide SiO_x layer).

intrinsic LPCVD of poly-Si films was deposited on both symmetrical lifetime test structures (textured and planar) and solar cell precursors (i.e., front-side textured, rear-side planar surfaces), followed by the phosphorus diffusion optimization process as mentioned above. The best results from the optimization process are highlighted in **Figure 13** and **Table 7**.

Comparing these results to the thick (~250 nm) thermal-SiO_x/poly-Si(n^+) passivated contacts (**Table 6**), the thin (~10 nm) thermal-SiO_x/poly-Si(n^+) passivated contacts on similar lifetime test structures (textured and planar) also exhibited excellent passivation qualities, attaining an implied-V_{OC} of 703 and 727 mV for the textured and planar case, respectively, after a symmetrical SiN_x capping step. Excellent film and diffusion uniformity was observed from the photoluminescence images; an example is shown in **Figure 13(c)** for the solar cell precursor structure (i.e., front-side textured, rear-side planar) with an electron-selective passivated contact being deposited on both sides (no SiN_x capping). However, it was observed that the absolute implied-V_{OC} values are slightly lower (few millivolts) than the thicker counterparts.

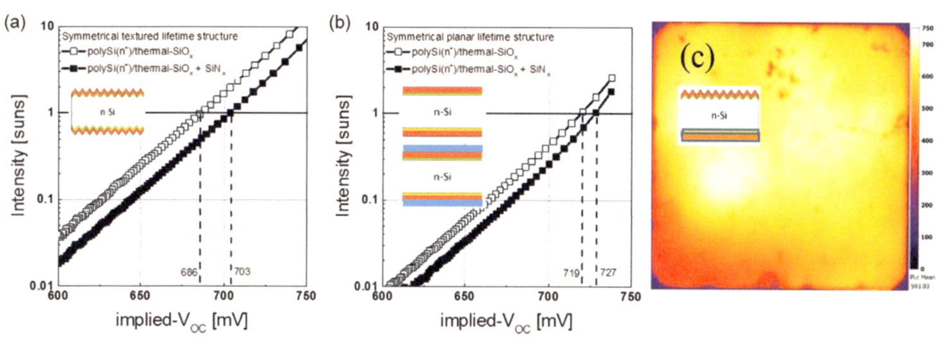

Figure 13.
Excellent passivation quality demonstrated from our in-house developed electron-selective (thermal-SiO_x/poly-Si(n^+)) passivated contact with thin (~10 nm) poly-Si(n^+) capping layers applied on both (a) symmetrical textured lifetime test structures, with iV_{OC} reaching 686 mV, and (b) symmetrical planar lifetime test structures, with iV_{OC} reaching ~720 mV, which further improves to 703 and 727 mV, respectively, after an additional standard SiN_x capping layer. Good film and doping uniformity can be observed from the PL images for both the symmetrical lifetime test structures and solar cell precursors (i.e., front-side textured, rear-side planar) as shown in (c).

Structure	Surface (poly-Si thickness)	SiN_x capped?	τ_{eff} (µs)	Total J_0 (fA cm^{-2})	iV_{OC} (mV)
A	Sym. planar (10 nm)	No	4229	16	719
A	Sym. planar (10 nm)	Yes	7277	10	727
B	Sym. textured (10 nm)	No	961	67	686
B	Sym. textured (10 nm)	Yes	1928	31	703
C	Asym. front txt., rear planar (10 nm)	No	2982	22	713
C	Asym. front txt, rear planar (10 nm)	Yes	6557	Inj. dep	741

Table 7.
Summary of the measured passivation quality parameters (τ_{eff}, total J_o, implied-V_{OC}) for an electron-selective passivated contact comprising an in situ thermal-SiO_x tunnel layer coupled with a thin (~10 nm) poly-Si(n^+) capping layer, evaluated on lifetime test structures which are symmetrically planar (structure A), symmetrically textured (structure B), and front-side textured and rear-side planar solar cell precursors (structure C).

To provide more insights, ECV measurements were performed on the thin poly-Si(n^+) layers on both the textured and planar surfaces and compared to the thick reference as shown in **Figure 14**. The following observations can be made: (i) the thin poly-Si(n^+) layer exhibits a higher phosphorus dopant concentration ($\sim 5 \times 10^{20}$ cm^{-3}) than the thicker counterpart ($\sim 2 \times 10^{20}$ cm^{-3}); and (ii) the poly-Si(n^+) layer on the textured surface exhibits a higher dopants in-diffusion than the planar surface, which could partially explain the lower measured implied-V_{OC} values for the former (i.e., 686 mV as compared to 719 mV).

Similar to the thick poly-Si(n^+) capped samples, an additional symmetrical SiN_x capping further enhances the overall passivation quality, such that the textured and planar lifetime structures now exhibit an improvement in the implied-V_{OC} by 17 and 8 mV, which is a relative improvement of 2.5 and 1.1%, respectively.

To summarize, we have demonstrated on a textured silicon surface the ability to obtain an excellently passivating SiN_x-capped electron-selective passivated contact (thermal-SiO_x/poly-Si(n^+)) with sufficiently thin poly-Si(n^+) thickness ($\sim$10 nm) to reduce the parasitic absorption issue while maintaining excellent passivation qualities (implied-V_{OC} values exceeding 700 mV). Re-optimizing the diffusion recipe for an ultrathin LPCVD of intrinsic poly-Si layer therefore solves the limitations encountered when using slow silicon etch technology, which limited the obtainable poly-Si(n^+) layer thickness to $\sim$70 nm (i.e., observing a drastic drop in passivation quality for thinner layers).

3.6 Simulation studies: reducing parasitic absorption from highly doped poly-Si layers

The excellent results from the earlier sections clearly demonstrate the potential of deploying double-sided passivated contacts for next-generation silicon solar cell

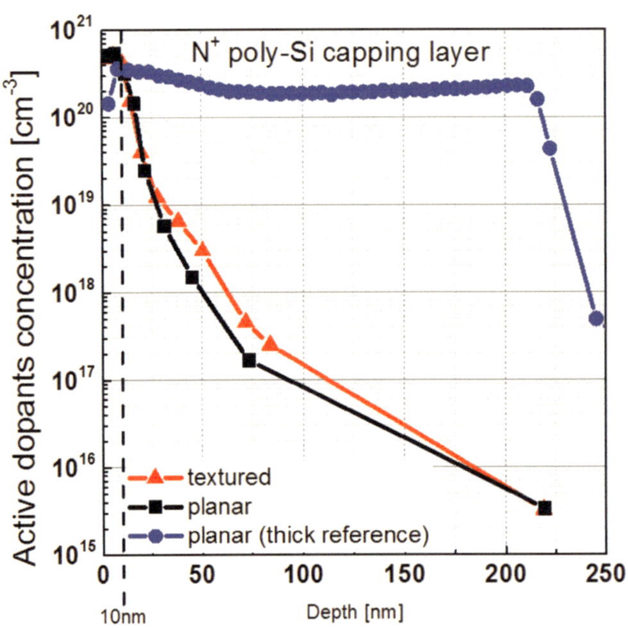

Figure 14.
Comparison of the ECV profile for ∼10-nm-thick poly-Si(n^+) layer on both planar and textured silicon surfaces. Additionally included is the ECV profile for a thick poly-Si(n^+) capping layer for comparison purpose. A higher dopant in-diffusion is observed for the textured surfaces, which partially explains the lower measured implied-V_{OC} values.

concepts, which in this work had been entirely realized on commercially available industrial tools. However, as mentioned earlier, one of the key issues if contact passivation is to be applied front-side also is to minimize parasitic absorption within the highly doped front-side poly-Si capping layer. Highly doped poly-Si is similar to transparent conductive oxide (TCO) layers deployed for silicon heterojunction solar cells, non-zero extinction coefficients, resulting in the inevitable parasitic absorption. This is even more pronounced, if applied front-side, thereby directly reducing the absorbable photogeneration current in the silicon wafer bulk (as the incident light is then first entering the parasitically absorbing poly-Si capping layer before entering the silicon wafer).

Hence, the objective of this section is to utilize an appropriate numerical calculation method to determine the parasitic absorption as a function of the (rear or front side) poly-Si capping layer thickness and then subsequently predict the corresponding solar cell efficiency potential of the correspondingly optimized passivated contact, being rear-side-only or front- and rear-side deployed in a solar cell. To address the above, the simulation program SunSolve™, available on PV Lighthouse [53], was utilized to study the impact of the doped poly-Si capping layer thickness on the maximum absorbable current density within the silicon wafer bulk $J_{absorbed,\ cell}$. Besides calculating $J_{absorbed,\ cell}$, the various optical losses can also be determined (i.e., front-reflected, front-escaped, rear-escaped, parasitic absorption in each layer, edge absorption) for the investigated solar cell precursors in this work.

To enhance the accuracy of the optical calculations, ellipsometry measurements were performed on all in-house fabricated samples, i.e., measuring our deployed dielectric films (SiN_x, SiO_x, AlO_x) as well as our optimized doped poly-Si capping layers, followed by a fitting and extraction of the wavelength-dependent optical refractive indices (n, k). These wavelength-dependent refractive indices were then imported into the SunSolve™ simulation program for a more realistic prediction of the current loss analysis, based on our own developed contact passivation layers. As an example, **Figure 15** shows the fitted wavelength-dependent refractive indices for the doped poly-Si layers in this work, which is further compared to the crystalline silicon reference [73]. As seen, the doped poly-Si layers do exhibit a higher extinction coefficient (k) compared to a c-Si reference within the visible to near-infrared region (400–900 nm). This again indicates that parasitic absorption is inevitable and should be minimized by thickness reduction while not compromising on the passivation quality. Further optimization work should also try to reduce the extinction coefficient of the poly-Si capping layers itself, i.e., by changing its chemical composition.

The current loss analysis results for a rear-side passivated contact solar cell (using SunSolve™) are shown in **Figure 16**. In order to account for internal back reflection, a local full-area metal contact scheme has been assumed (see **Figure 16**) (this can be realized by local laser ablation, forming contact openings in the SiN_x passivation layer and a subsequent full-area metallization).

It can be seen that for a solar cell with a conventional front-side boron-diffused junction and a rear-side electron-selective passivated contact (thermal-SiO_x/poly-Si(n^+)), the parasitic absorption arising from the rear-side poly-Si(n^+) capping layer can be directly addressed by reducing the rear-side poly-Si film thickness (e.g., the parasitic absorption current loss reduces from 0.55 mA cm^{-2} for a 250-nm-thick poly-Si(n^+) layer to 0.02 mA cm^{-2} for a 10-nm-thick poly-Si(n^+) layer). The reduction in the parasitic absorption directly enhances the potentially absorbable current in the wafer bulk ($J_{absorbed,\ cell}$), which in this case improves from ~40.7 to ~41.1 mA cm^{-2}.

Interestingly, it was observed that for a poly-Si capping layer thickness lower than 25 nm, the $J_{absorbed,\ cell}$ saturates at ~41.1 mA cm^{-2}. On hindsight, we would

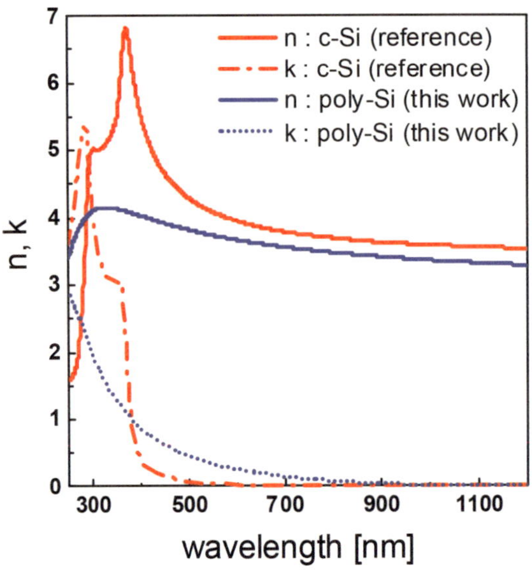

Figure 15.
Optical indices extracted for the doped poly-Si layers in this work, based on ellipsometry measurements and its subsequent fitting by the Tauc-Lorentz model. Also included is the crystalline silicon optical index data for reference. The poly-Si films do exhibit higher extinction coefficient values than the c-Si wafer bulk within the visible to near-infrared regions (400–900 nm), clearly indicating the need to optimize the poly-Si capping layer thickness in order to reduce parasitic absorption.

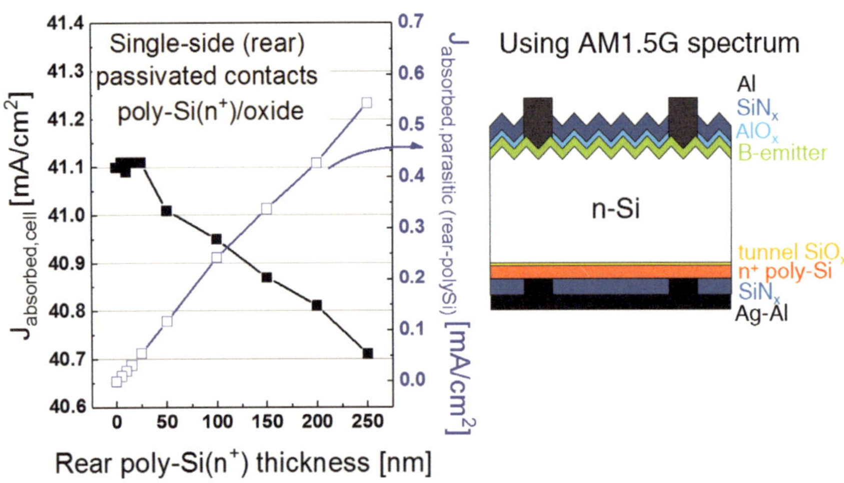

Figure 16.
Numerically calculated absorbed photogeneration current in the silicon solar cell bulk ($J_{absorbed,\ cell}$) and the parasitic absorption contributed by the rear-side poly-Si(n^+) capping layer ($J_{absorbed,\ parasitic\ (rear\text{-}polySi)}$), as a function of its thickness from 250 nm down to 0 nm, for a rear-side passivated contact solar cell, adopting a conventional front-side boron-diffused emitter junction, and the investigated rear-side electron-selective passivated contacts (tunnel oxide/poly-Si(n^+)). Reducing the rear-side poly-Si(n^+) layer thickness leads to a significant reduction on parasitic absorption (up to 0.55 mA cm^{-2}) and a corresponding gain in the photogeneration current $J_{absorbed,\ cell}$ (up to 0.4 mA cm^{-2}). Interestingly, $J_{absorbed,\ cell}$ saturates for a poly-Si(n^+) layer thickness lower than 25 nm, despite a further reduction of rear-side parasitic absorption (see text).

expect that as the poly-Si capping layer thickness reduces, photons which were not absorbed in the first pass within the cell bulk would now have an increased probability of being parasitically absorbed at the rear-side metal contacts. Indeed, the numerical calculations confirm that hypothesis in which the calculated parasitic absorption within the rear-side metal contacts increases from ~0.73 mA cm^{-2} with a 250-nm-thick poly-Si to ~0.78 mA cm^{-2} with a 10-nm-thick poly-Si layer. Additionally, there was also a clear increasing trend in the front-escaped current density from ~1.91 mA cm^{-2} with a 250-nm-thick poly-Si to ~2 mA cm^{-2} with a 10-nm-thick poly-Si layer. These two effects were found to limit the potential $J_{\text{absorbed, cell}}$ in case of deploying very thin rear-side poly-Si capping layers.

Hence, for the purpose of device integration, our numerical findings suggest that when considering tunnel oxide/poly-Si(doped) passivated contacts at the rear surface, it would suffice to shrink down the rear-side poly-Si thickness to 25 nm (thinner layers will not further improve the photogeneration current $J_{\text{absorbed, cell}}$ within the silicon wafer). However, a thicker rear-side poly-Si layer may be more suited to accommodate screen-printed, industrial fire-through metal contacts, without damaging the interface passivation (see the next section). Hence, a trade-off of between these two requirements is needed and to be investigated in the future work.

Figure 17 presents a pie chart summary for the current loss analysis of the simulated solar cell structure shown in **Figure 16**, which adopts a rear-side poly-Si(n$^+$) capping layer with an experimentally realizable thickness of 10 nm as mentioned in the earlier sections. Based on this single-sided (rear-side) passivated contact solar cell structure, the parasitic absorption contribution by the rear-side poly-Si(n$^+$) layer leads to a negligible low 0.04% of the total AM1.5G incident current density of 46.32 mA cm^{-2}, amounting to 0.02 mA cm^{-2} only. The bulk of the incident photon current density is absorbed by the silicon wafer (88.72%), although this could be further enhanced when better front-side anti-reflection coatings are available for deployment (currently, a front-reflected current density loss of 4.66% is calculated for our in-house deployed thin-film AlO_x/SiN_x anti-reflection stack). The second highest current loss channel is the front-escaped current density at 4.32%. Please note that this loss channel cannot be reduced: Photons which are desired to enter the silicon wafer will also be able to leave it. Actually, the higher the percentage loss due to front surface escape, the better the optical performance of the solar cell. The metal grid at the front and rear accounts for a total current loss of 2.05% based on our in-house available screen designs. Taking all optical current losses into account, the maximum absorbable photon current density in the silicon wafer is ~41.1 mA cm^{-2}.

Extending the analysis from a solar cell with a single rear-side-only passivated contact toward double-sided passivated contacts, the same current loss analysis

%	Photocurrent
4.66	2.16 mA cm^2
4.32	2.00 mA cm^2
0.38	1.76 mA cm^2
0.07	0.03 mA cm^2
0.00	0 mA cm^2
88.72	41.10 mA cm^2
0.00	0 mA cm^2
0.04	0.02 mA cm^2
0.00	0 mA cm^2
1.67	0.77 mA cm^2
0.09	0.04 mA cm^2
0.04	0.02 mA cm^2

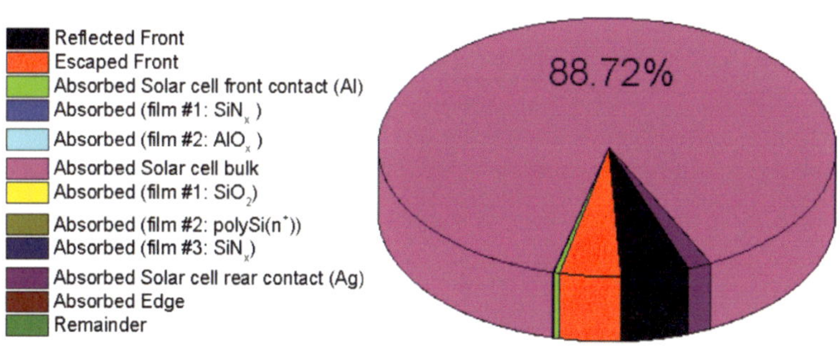

Figure 17.
Pie chart representing the SunSolve™ current loss analysis for the solar cell structure sketched in **Figure 16**, *featuring an experimentally realizable rear-side poly-Si(n$^+$) capping layer thickness of 10 nm. In this case, the rear-side poly-Si(n$^+$) capping layer exhibits a negligible parasitic absorption of 0.04% at 0.02 mA cm^{-2}, out of the total incident current density of 46.32 mA cm^{-2} for the utilized AM1.5G solar spectrum. For a rear-side-only passivated contact solar cell, the maximum absorbable photocurrent density in the silicon wafer is ~41.1 mA cm^{-2} (88.72% of the incoming solar spectrum).*

approach was applied to front- and rear-side passivated contact solar cells, exhibiting an optically negligible rear-side capping layer thickness of 3 nm, as experimentally realized. As sketched in **Figure 18**, this solar cell structure consists of a front-side textured surface with our developed electron-selective (tunnel oxide/poly-Si(n$^+$)) passivated contacts, and a rear-side planar surface with our developed hole-selective (tunnel oxide/poly-Si(p$^+$)) passivated contacts. This is followed by the standard dielectric coatings (SiO_x, SiN_x, AlO_x) at both surfaces to serve both passivation and anti-reflection purposes, prior to the screen-printed fire-through metal contacts at both sides. Adopting an experimentally realizable rear-side poly-Si(p$^+$) capping layer thickness of 3 nm (see earlier section), the influence of the front-side poly-Si capping layer thickness on the $J_{absorbed,\ cell}$ is investigated. **Figure 18** shows that the parasitic absorption by the front-side poly-Si capping layer has a much more severe and significant impact on the remaining absorbable current density in the solar cell bulk ($J_{absorbed,\ cell}$). If contact passivation is applied front-side, $J_{absorbed,\ parasitic\ (front\ poly\text{-}Si)}$ is as high as ~20.8 mA cm^{-2} for a 250-nm-thick poly-Si(n$^+$) layer, and it reduces to ~1 mA cm^{-2} for a 5-nm-thick poly-Si(n$^+$) layer. Front-side poly-Si layer thickness reduction therefore directly translates into a significant gain in $J_{absorbed,\ cell}$, approximately by the same amount (i.e., increasing from ~21 to ~40.3 mA cm^{-2}).

Accordingly, the pie chart current loss analysis for the double-sided passivated contact solar cell structure depicted in **Figure 3** is shown in **Figure 19** for the case of an experimentally realizable front-side poly-Si(n$^+$) capping layer thickness of 10 nm and an experimentally realizable rear-side poly-Si(p$^+$) capping layer thickness of 3 nm. **Figure 19** shows that the presence of the 10-nm-thick front-side poly-Si(n$^+$) capping layer contributes to a comparatively higher parasitic absorption loss (4.32%) than the rear-side poly-Si(p$^+$) capping layer (0.01%), based on a total incident current density of 46.32 mA cm^{-2} (AM1.5G spectrum). The remaining potentially absorbable current density within the solar cell bulk stands at ~85.56% (~39.6 mA cm^{-2}), which is ~3.16% lower than a rear-side-only passivated contact scheme. Hence, it is clear that although double-sided passivated contact solar cells could deliver excellent passivation on both sides of the wafer (thereby reaching higher open-circuit voltages V_{OC} than rear-side-only passivated contact solar cells or conventional diffused solar cells), there is still a trade-off with increased front-side parasitic absorption, demanding more optimization efforts.

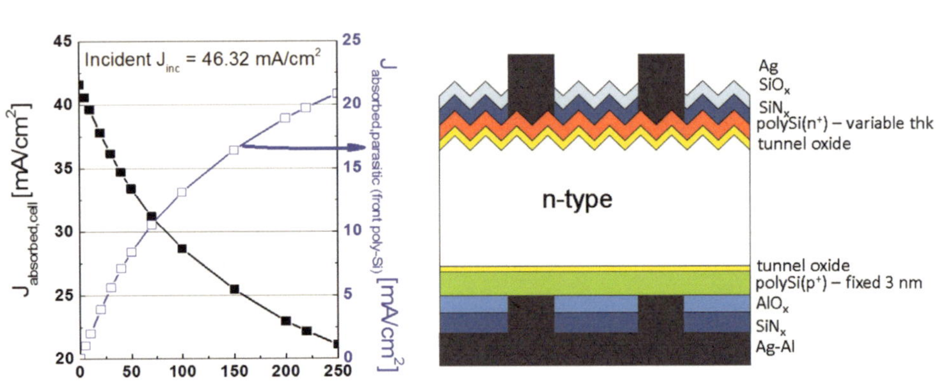

Figure 18.
Numerically calculated photon current absorption for a double-sided passivated contact solar cell. The rear-side hole-selective poly-Si(p$^+$) capping layer thickness is fixed at 3 nm, while the front-side electron-selective poly-Si(n$^+$) capping layer thickness is varied from 0 nm to 250 nm. Front-side parasitic within the poly-Si(n$^+$) capping layer ($J_{absorbed,\ parasitic\ (front\ poly\text{-}Si)}$) has a severe impact on the absorbable photon current density within the silicon wafer ($J_{absorbed,\ cell}$).

%	Photocurrent	
3.37	0.16 mA cm^{-2}	Reflected Front
4.76	0.22 mA cm^{-2}	Escaped Front
0.13	0.01 mA cm^{-2}	Absorbed Solar cell front contact (Ag)
0.00	0.00 mA cm^{-2}	Absorbed (film #1: SiOx)
0.06	0.03 mA cm^{-2}	Absorbed (film #2: R2 SiNx)
4.32	0.20 mA cm^{-2}	Absorbed (film #3: front poly-Si(n+))
0.00	0.00 mA cm^{-2}	Absorbed (film #4: tunnel SiO2)
85.56	39.63 mA cm^{-2}	Absorbed Solar cell bulk (layer #2)
0.00	0.00 mA cm^{-2}	Absorbed (film #1: SiO2)
0.01	0.00 mA cm^{-2}	Absorbed (film #2: rear poly-Si(p+))
0.00	0.00 mA cm^{-2}	Absorbed (film #3: AlOx)
0.00	0.00 mA cm^{-2}	Absorbed (film #4: R2 SiNx)
1.65	0.76 mA cm^{-2}	Absorbed Solar cell rear contact (Al)
0.10	0.00 mA cm^{-2}	Absorbed Edge
0.03	0.00 mA cm^{-2}	Remainder

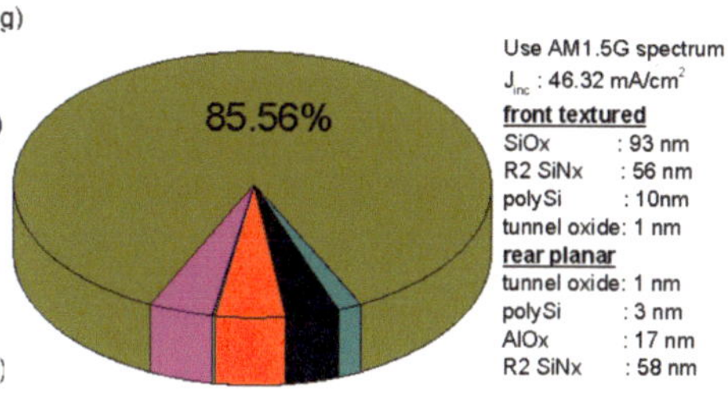

Figure 19.
Pie chart representing the SunSolve™ current loss analysis for the double-sided passivated contact solar cell structure sketched in ***Figure 18****, featuring a rear-side poly-Si(p$^+$) capping layer thickness of 3 nm and a front-side poly-Si(n$^+$) capping layer thickness of 10 nm. The 10-nm-thin front-side poly-Si(n$^+$) capping layer still contributes to the parasitic absorption (4.32% at 0.2 mA cm^{-2}), whereas parasitic absorption within the 3-nm rear-side poly-Si(p$^+$) capping layer can be neglected, based on a total incident current density of 46.32 mA cm^{-2} for the utilized AM1.5G solar spectrum. For the double-sided passivated contact solar cell, the maximum absorbable photocurrent density in the silicon wafer is now ~39.6 mA cm^{-2} (85.56% of the incoming solar spectrum).*

3.7 Compatibility of screen printing (using conventional screen-printing pastes) on our developed passivated contact layers

In earlier sections, the feasibility of the electron-selective and hole-selective passivated contacts has been demonstrated, both on symmetrical lifetime test structures and asymmetrical solar cell precursors as sketched in **Figure 11** (in the as-deposited state and after an additional symmetrical SiN_x capping). The remaining solar cell fabrication step would be the formation of metal contacts toward these thin-film passivated contacts, without damaging the passivation quality underneath these contacts. As a first attempt, conventional metal contacting schemes, i.e., screen printing, as commonly deployed for conventional silicon solar cells (exhibiting double-sided diffused junctions), were performed on our lifetime and solar cell precursors. In particular, we tested our industrial in-house fire-through and non-fire-through screen-printing pastes, based on Ag, Ag/Al, or Al material formulations. The corresponding results were compared to a nonindustrial research reference contact, deploying thermally evaporated Ag contacts.

In summary, so far, using conventional screen-printing pastes, screen printing works only on comparatively thick poly-Si(n$^+$) layers, i.e., requiring a poly-Si(n$^+$) thickness of 150 nm or larger. So far, it does not work on poly-Si(p$^+$) layers. The SEM results presented in **Figure 20** sum up these observations.

(I) A fire-through Ag paste (as conventionally used to contact n-doped silicon material) is able to contact our standard ~250-nm-thick poly-Si(n$^+$) layers conformably, without any issues, i.e., exhibiting a low contact resistance (13 mΩ cm^2) and no void issues or punch-through effects underneath the contact (see **Figure 20** (top, left)). The investigated fire-through Ag paste is suitable for rear-side contacting poly-Si(n$^+$) capping layers down to a thickness of 150 nm; however, it fails to contact our ultrathin ~10-nm poly-Si(n$^+$) capping layer, as outlined in some more detail later.

Deploying industrial screen printing for rear-side-only passivated contact solar cells, we currently reach a solar cell efficiency of 21.7%, using our 250-nm "standard" rear-side SiO_x/poly-Si(n$^+$) contact passivation layers (see **Figure 21** and **Table 8**).

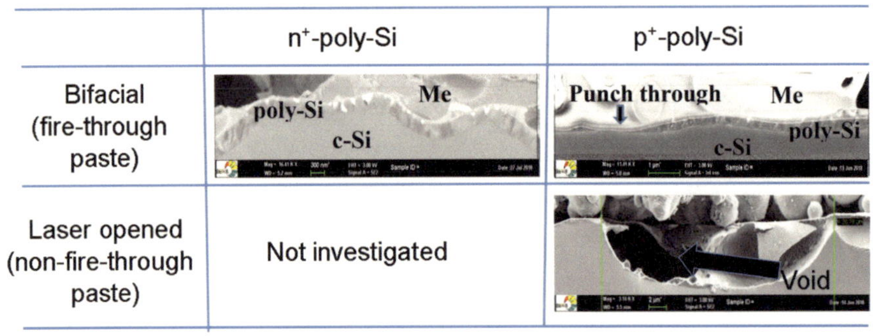

Figure 20.
SEM images taken for poly-Si(n^+) and poly-Si(p^+) layers contacted via conventional screen printing, using various commercially available pastes: (i) bifacial fire-through pastes, i.e., Ag paste for contacting n-doped Si and Al paste for contacting p-doped Si, and (ii) non-fire-through Ag/Al pastes, using laser ablation to form local contact openings prior to screen printing. Screen printing works only in case of contacting moderately thick (150–250 nm) electron-extracting poly-Si(n^+) capping layers. In all other cases, issues like void formation or a local "punch through" of the metal paste (locally contacting the c-Si wafer instead of the poly-Si capping layer) occur.

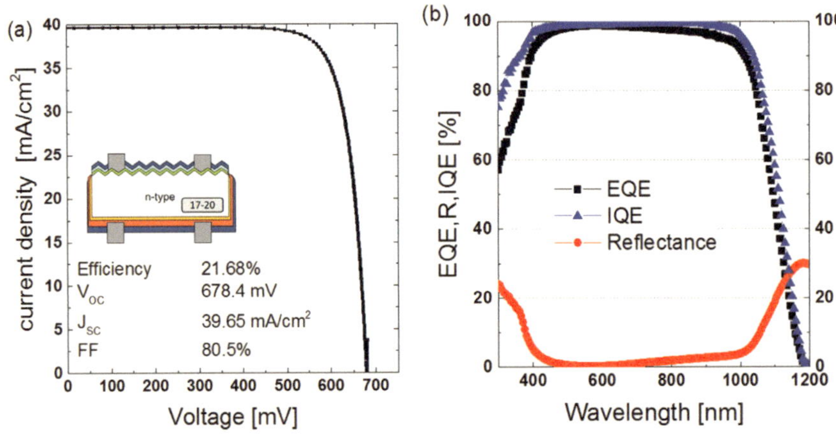

Figure 21.
(a) Measured I–V curve of our current rear-side passivated contact record efficiency cell, exhibiting a rear-side wet-chemically formed SiO_x tunnel layer, a 250-nm poly-Si(n^+) capping layer, and a conventionally formed (boron-diffused, AlO_x/SiN_x passivated and screen printed) front-side contact. (b) The correspondingly measured external/internal quantum efficiency, EQE, IQE, and the measured reflectance for the same cell.

Reducing the rear-side poly-Si(n^+) capping layer thickness (separate batch, hereby only reaching 21.3% for the solar cell with the 250-nm-thick poly-Si reference layer), we were able to observe a clear increase in short-circuit current density (see **Table 9**). By thinning down the rear-side poly-Si(n^+) capping layer thickness from 250 nm down to 150 nm, using etch-back technology, we gain ~0.4 mA cm^{-2} in short-circuit current density, reaching again a best cell efficiency of 21.7%. Up to a thickness of 150 nm, the poly-Si(n^+) thinning did neither significantly affect the open-circuit voltage V_{oc} nor the fill factor of the solar cell (compare **Table 8**). However, the samples with a 100-nm rear-side poly-Si capping layer exhibit a drop in V_{oc} (~15 mV). This resulted from a local punch through of the screen-printed metal paste, similar to the SEM image as shown in **Figure 20** (top, right).

(II) A fire-through Ag/Al paste (as conventionally used to contact p-doped silicon material) could not contact our standard ~250-nm-thick poly-Si(p^+) layers properly: There are several regions where the paste is observed to consume the poly-Si(p^+) layer, causing a thinning of the poly-Si(p^+) layer and some local

Cell type: thickness of rear poly-Si(n^+)	V_{oc} (mV)	J_{sc} (mA cm^{-2})	*FF* (%)	*Eff* (%)
250 nm	678	39.7	80.5	21.7

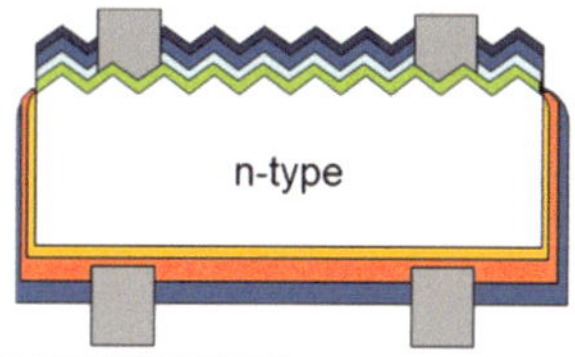

Table 8.
I–V data of the rear-side passivated contact record cell, deploying conventional bifacial screen printing for metallization.

Cell type: thickness of rear poly-Si(n^+)	V_{oc} (mV)	J_{sc} (mA cm^{-2})	*FF* (%)	*Eff* (%)
250 nm	672	39.4	80.2	21.3
150 nm	676	39.8	80.7	21.7
100 nm	660	39.6	80.2	21.0

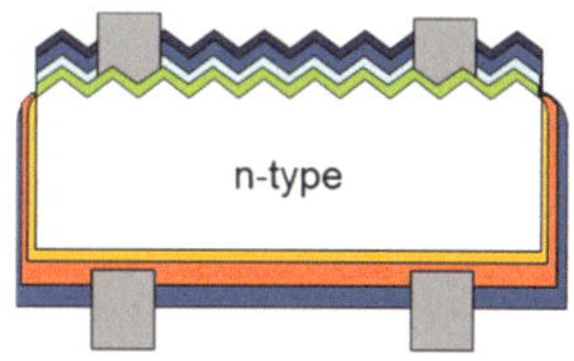

Table 9.
I–V data of rear-side passivated contact solar cells, with a varying rear-side poly-Si(n^+) capping layer thickness.

"punch-through" areas (see **Figure 20** (top, right)). This in turn leads to local shunting (in case of using an n-type wafer) and to a severe degradation of contact passivation quality, as evident from the final measured cell V_{oc} values.

This issue can be likely attributed to the presence of the Al alloy within the paste, which is typically responsible for forming the back surface field regions in conventional silicon solar cells. Al alloying is known to partially consume crystalline silicon material: thus, our thin poly-Si(p^+) capping layers will be consumed upon contact firing of the screen-printed Ag/Al paste, leading to the just outlined local "punch-through" effects.

(III) A non-fire-through pure Al paste (as conventionally used to contact a p-doped silicon wafer in order to form locally Al-alloyed back-surface-field (BSF) regions within the wafer) was found to create large voids in several regions (see **Figure 20** (bottom, right)) and to consume the entire poly-Si(p^+) passivated contacts, leading to a drastic drop in contact passivation quality and measured device performance.

It is possible to use femtosecond laser ablation, in order to create damage-free local contact openings (i.e., locally ablating the overlaying SiN_x layer without damaging the underlying poly-Si(p^+) capping layer). Using a femtosecond laser at an ultraviolet wavelength of 330 nm, the onset of laser fluence for optimized SiN_x ablation is 0.08 J cm^{-2}. Within the optimized process window, the lifetime is preserved after laser ablation (as indicated by photoluminescence imaging), and the SiN_x is fully ablated (as indicated by optical microscope imaging) (see **Figure 22**). However, the paste composition of the screen-printing paste has to be altered, in order to enable a subsequent damage-free contacting of our (thick or ultrathin) poly-Si(p^+) layers. Corresponding research activities, in cooperation with a paste manufacturer, are currently initiated.

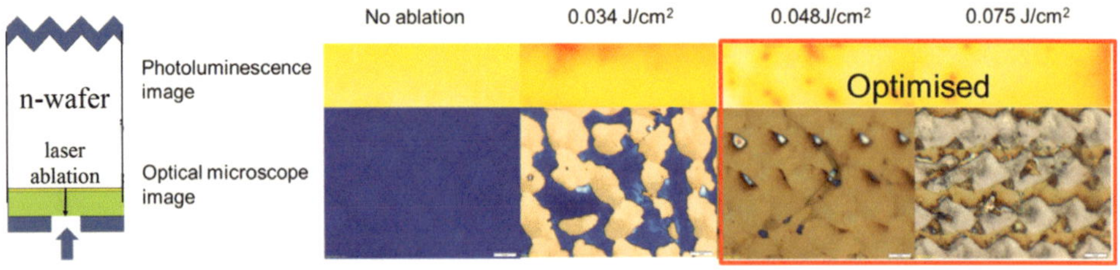

Figure 22.
(Left) Sketch of the laser process to locally laser ablate SiN_x on top of thin-film poly-Si, in order to form local contact openings for further metallization, aiming at conventional screen printing, using a non-fire-through paste. (Right) Photoluminescence images and optical microscope images inside the opening, taken for a screened range of laser fluence.

(IV) As expected, our research reference thermal evaporated Ag contacts were able to form damage-free conformal low resistivity contacts to our developed SiO_x/poly-Si(p^+) and ALD-AlO_x/poly-Si(p^+) passivated contacts, thereby enabling a nonindustrial full-area reference contact on hole-extracting poly-Si(p^+) capping layers [34].

As just outlined above, using our conventional screen-printing metal pastes and fast-firing conditions, thus far we were not able to successfully contact hole-extracting poly-Si(p^+) layers as well as ultrathin 10-nm electron-extracting poly-Si(n^+) layers. Thus, a closer attention toward (i) an optimization of the metal paste itself, i.e., tuning its chemical composition, and (ii) an optimization of the fast-firing conditions, applied after screen printing, in order to form a low resistivity contact, is necessary.

To address the latter, an asymmetric lifetime test structure, featuring a textured front surface and a planar rear surface, symmetrically passivated contact by our ultrathin (~10 nm) electron-selective, thermal-SiO_x/poly-Si(n^+) passivated contact layers, was utilized. The passivation quality of these samples in the as-deposited state was measured first, followed by a symmetrical deposition of the passivation/anti-reflective SiN_x film, and its passivation quality was remeasured. Then, the samples were subjected to different fast-firing peak temperatures (650, 660, 680, 700, 720, 740, 760°C), thereby mimicking different fast-firing conditions after screen printing, and the resulting final passivation quality was remeasured again (see **Figure 23**).

For our ultrathin 10-nm SiO_x/poly-Si(n^+) contact passivation layers, a severe degradation of passivation quality after fast-firing is observed (see **Figure 23**). In the as-deposited state, our asymmetrical lifetime test structures with electron-selective passivated contacts were exhibiting good passivation quality with average $\tau_{eff}/J_0/iV_{oc}$ values of ~3.3 ms /35.4 fA cm^{-2}/704 mV. Upon the subsequent symmetrical SiN_x capping, the passivation quality got further enhanced, i.e., reaching excellent average $\tau_{eff}/J_0/iV_{oc}$ values of ~7.3 ms /14.4 fA cm^{-2} /720 mV. However, after an additional short high-temperature treatment, in case of using our ultrathin 10-nm electron-extracting contact passivation layers, the passivation quality drops significantly. For example, if we adopt a fast-firing peak temperature of 740°C (which is currently utilized for our conventional double-sided diffused silicon solar cells), a drastic drop in passivation quality occurs, in which the $\tau_{eff}/J_0/iV_{oc}$ values degrade to ~1.2 ms/91 fA cm^{-2}/684 mV. This effect is less severe, but still significant, if lower fast-firing peak temperatures can be deployed (see **Figure 23**). It seems like our current ultrathin electron-selective passivation layers are not firing stable, especially if deploying high peak firing temperatures (they still do outperform conventionally diffused front-side contacts, though). Interestingly, this

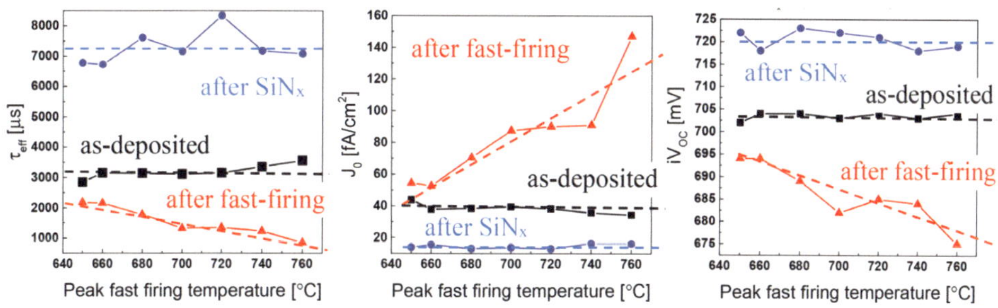

Figure 23.
Comparison of the measured passivation quality for the asymmetrical lifetime test structures comprising a textured front and a planar rear surface, passivated with electron-extracting thermal-SiO_x/10 nm-poly-Si(n^+) passivated contacts, in the as deposited state, after additional SiN_x deposition and after a fast-firing belt furnace temperature treatment. The straight lines are a guide to the eyes.

is not the case for the 250-nm-thick "standard" layers. ECV measurements confirm that after fast-firing, the dopants within the poly-Si(n^+) capping layer have out-diffused into the silicon wafer bulk, thereby effectively reducing field-effect passivation and thus the observed lifetimes of the samples. More detailed investigations are currently ongoing.

Thus, more efforts to render our ultrathin contact passivation layers firing stable, i.e., by deploying lower peak firing temperatures and/or changing the chemical composition of the ultrathin LPCVD of poly-Si capping layers, are necessary. Furthermore, efforts to optimize the composition of the screen-printing paste itself, in order to be able to successfully contact ultrathin poly-Si layers using screen printing, will be undertaken. An alternative work plan is to investigate low-temperature inline plating, as a possible approach to contact our ultrathin SiO_x/poly-Si contact passivation layers.

3.8 Cell efficiency potential prediction: single-sided versus double-sided contact passivation

As already indicated in the introduction part, we can determine a practical solar cell efficiency potential of our investigated solar cell structures, adopting either a rear-side-only passivated contact scheme or a double-sided passivated contact scheme. Using Brendel's model [54], and explicitly considering measured front-side contact resistance and contact recombination parameters (i.e., the combined front-side saturation current density $J_{0,\ \mathrm{front}}$, combing the contributions from both the non-metallized/passivated regions $J_{0,\ \mathrm{non\text{-}metal}}$ and from the metallized regions $J_{0,\ \mathrm{metal}}$), it is possible to calculate a practical solar cell efficiency potential as a function of the rear-side passivated contact layer properties, i.e., the rear-side recombination current density $J_{0,\ \mathrm{rear}}$ and the rear-side contact resistance $R_{c,\ \mathrm{rear}}$ of the rear-side passivated solar cell contact. By fixing the front-side $J_{0,\ \mathrm{front}}$ and $R_{c,\ \mathrm{front}}$ contributions, iso-efficiency contour plots can be calculated as a function of the rear-side $J_{0,\ \mathrm{rear}}$ and the rear-side $R_{c,\ \mathrm{rear}}$ (thereby generalizing Brendel's model [54]). The goal of the cell efficiency prediction is twofold: (1) to determine if adopting a double-sided passivated contacts scheme is better than the single-sided (rear) passivated contact scheme and (2) to determine if a full-area rear-side contacting scheme is better than a bifacial contacting scheme.

Firstly, **Figure 24** shows a comparison of solar cells with a rear-side-only passivated contact scheme, comprising a conventional front-side textured surface with a boron-diffused emitter, passivated by a standard AlO_x/SiN_x double-layer

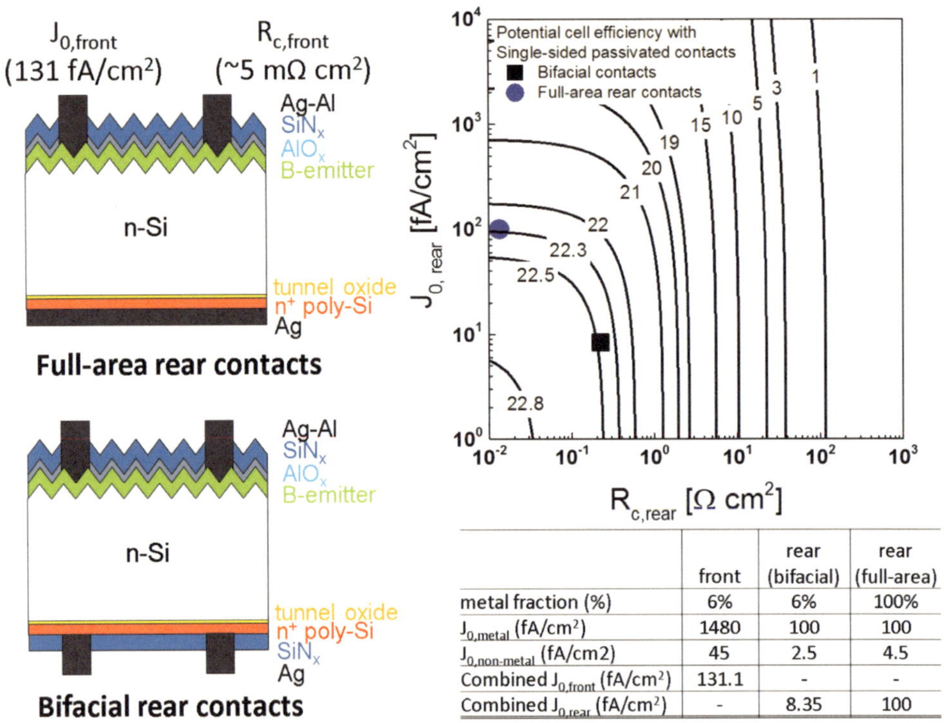

	front	rear (bifacial)	rear (full-area)
metal fraction (%)	6%	6%	100%
$J_{0,metal}$ (fA/cm²)	1480	100	100
$J_{0,non\text{-}metal}$ (fA/cm2)	45	2.5	4.5
Combined $J_{0,front}$ (fA/cm²)	131.1	-	-
Combined $J_{0,rear}$ (fA/cm²)	-	8.35	100

Figure 24.
Practical solar cell efficiency potential for a rear-side-only passivated contact solar cell (as a function of the quality of the rear-side passivated contact, i.e., its recombination current density $J_{0,\ rear}$ and its contact resistance $R_{c,\ rear}$), adopting a conventional front-side boron-diffused emitter and a rear-side electron-selective SiO_x/poly-Si(n^+) passivated contact, realized either in a full-area contact configuration or in a bifacial contact configuration. The measured current recombination densities $J_{0,\ rear}$ and the correspondingly measured contact resistances $R_{c,\ rear}$ of our developed rear-side SiO_x/poly-Si(n^+) electron-extracting passivated contacts are inserted within the iso-efficiency plot (blue dot, full-area contact; black square, bifacial contact). The corresponding practical solar cell efficiency potential using our developed SiO_x/poly-Si(n^+) passivated contacts is 22.3%, if a full-area rear-side contact is deployed, and 22.5%, if a bifacial contact is deployed.

anti-reflection coating and metallized by conventional screen printing (using a fire-through Ag-Al paste). The rear side composes of our developed electron-selective passivated contacts (thermal-SiO_x/poly-Si(n^+)), utilizing an experimentally achievable ~10-nm-thick poly-Si(n^+) layer and either a full-area Ag contact or a bifacial Ag contact with a contact area fraction similar to the front side (6%). For the efficiency potential prediction, a conservative, industrial feasible $J_{0,\ front}$ value of 131 fA cm^{-2} and $R_{c,\ front}$ value of ~5 mΩ cm^2 have been used. This corresponds to a $J_{0,\ front,\ pass}$ value of 45 fA cm^{-2} underneath the AlO_x/SiN_x passivated B-diffused regions [74] and a $J_{0,\ front,\ metal}$ value of 1480 fA cm^{-2} underneath the metal contacts [2, 75, 76], assuming a front-side metal contact area fraction of 6%.

Regarding our developed rear-side SiO_x/poly-Si(n^+) contact passivation layers, the corresponding properties have been measured explicitly: Utilizing the symmetrical planar lifetime test structures with electron-selective passivated contacts discussed in earlier sections, the single-sided $J_{0,\ rear,\ pass}$ values for the as-deposited and for the additionally SiN_x-capped samples were measured as 4.5 and 2.5 fA cm^{-2}, respectively. The recombination current density underneath the metal contact $J_{0,\ rear,\ metal}$ has been determined separately as ~100 fA cm^{-2}, using intensity-dependent PL imaging and our in-house developed Griddler software [77]. Please note that metal contact recombination after industrial screen printing is significantly reduced (more than one order of magnitude) if deploying contact passivation (i.e., comparing a $J_{0,\ front,\ metal}$ value of 1480 fA cm^{-2} to a $J_{0,\ rear,\ metal}$ value of

~100 fA cm^{-2}). Thus, a combined $J_{0,\ rear}$ value can be determined to be 8.35 and 100 fA cm^{-2} in case of a rear-side bifacial contact scheme or a full-area rear-side contact scheme, respectively (see **Figure 24**).

As discussed in Brendel's paper [54], in case of a rear-side bifacial contact, the recombination current density $J_{0,\ rear}$ scales with the rear-side contact area fraction, whereas the effective rear-side contact resistance $R_{c,\ rear}$ scales inversely with the rear-side contact area fraction. Again, regarding our developed rear-side SiO_x/poly-Si(n^+) contact passivation layers, the contact resistance $R_{c,\ rear}$ of our developed tunnel layer passivated contact has been measured explicitly: Based on our dark I–V test structures, as described in the introduction part of this paper and outlined in **Figure 1(f)**, the values for the bifacial and full-area rear-side contacts were measured to be 0.22 Ω cm^2 and 13.3 mΩ cm^2, respectively. These $J_{0,\ rear}$ and $R_{c,\ rear}$ values were then inserted into our calculated iso-efficiency contour plot in **Figure 24**, allowing a realistic prediction of the efficiency potential for a rear-side passivated contact solar cell in a bifacial or full-area configuration: As can be seen, the practical solar cell efficiency potential of a solar cell, adopting a conventional front-side boron-diffused emitter and a simple full-area rear-side passivated contact, is 22.3%. In case a bifacial contact is deployed, the practical solar cell efficiency potential is 22.5%. Using a rear-side bifacial contact instead of a full-area rear-side contact can therefore slightly enhance the solar cell efficiency by a relative gain of 0.9%.

The corresponding calculation of the practical efficiency potential for double-sided passivated contact solar cells is shown in **Figure 25**. As discussed in earlier sections, this solar cell concept features an optimized solar cell architecture considering our experimental finding, i.e., featuring a textured front surface with an electron-selective passivated contact (thermal-SiO_x/poly-Si(n^+)) and a planar rear surface with a hole-selective passivated contact (thermal-SiO_x/poly-Si(p^+)). The front surface of these cells is capped by a double-layer anti-reflection/passivation coating (SiO_x/SiN_x) and assumed to be contacted via screen-printed fire-through Ag contacts. The rear-side hole-selective passivated contacts are assumed to be either contacted by a full-area Ag contact or to be capped by a double-layer anti-reflection/passivation coating (AlO_x/SiN_x), forming a screen-printed bifacial fire-through Ag-Al contact.

Accordingly, in order to equate the rear-side $J_{0,\ rear}$ and $R_{c,\ rear}$ values for these two different contact schemes, we apply measured values, and we then plot the practical efficiency potential as a function of the quality of the front-side passivated contact ($J_{0,\ front}$ and $R_{c,\ front}$). The rear-side $J_{0,\ rear}$ value underneath the hole-selective passivated contact region was determined from the symmetrical lifetime test structures mentioned in earlier sections, while the rear-side $R_{c,\ rear}$ value was determined using the dark I–V test structures sketched in **Figure 1(f)** for the full-area case (using thermal evaporated Ag instead of screen printed Ag) and correspondingly inversely scaled with the contact-area fraction in case of the bifacial contact. It is to be noted that for our developed poly-Si(p^+) capping layers, the conventional screen-printing pastes were observed to consume the relatively thin poly-Si capping layer, thereby significantly degrading the rear-side $J_{0,\ metal}$ and $R_{c,\ rear}$ values (as reported in the former section). Nonetheless, in order to predict the practical efficiency potential of double-sided passivated contact solar cells, we assume this problem to be solved, i.e., we assume that applying a screen-printed contact on a hole-extracting poly-Si(p^+) capping layer will degrade our measured contact properties only in the same way as we observe it in case of an electron-extracting contact. Thus, as a first order of approximation, we assume the same $J_{0,\ metal}$ values for a metal contacting the hole-selective passivated contact as we measured it in case of a 250-nm-thick screen-printed SiO_x/poly-Si(n^+)

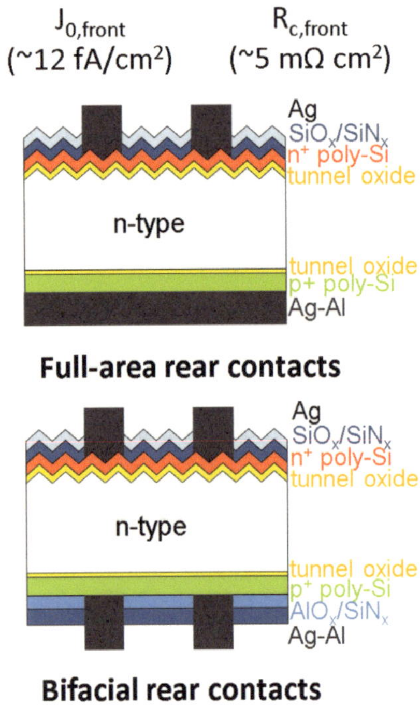

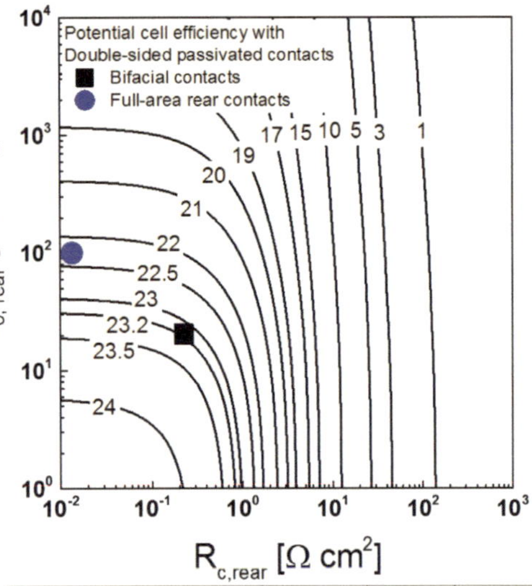

	front	rear (bifacial)	rear (full-area)
metal fraction (%)	6%	6%	100%
$J_{0,metal}$ (fA/cm²)	100	100	100
$J_{0,non\text{-}metal}$ (fA/cm2)	6.65	15.4	19.5
Combined $J_{0,front}$ (fA/cm²)	12.25	-	-
Combined $J_{0,rear}$ (fA/cm²)	-	20.48	100

Figure 25.
Practical solar cell efficiency potential for a double-sided passivated contact solar cell (as a function of the quality of the rear-side hole-extracting passivated contact, i.e., its recombination current density $J_{0,\ rear}$ and its contact resistance $R_{c,\ rear}$), adopting an ultrathin 10-nm electron-selective SiO_x/poly-Si(n^+) passivated contact on the textured front-side and a hole-selective SiO_x/poly-Si(p^+) passivated contact on the planar rear-side, realized either in a full-area contact configuration or in a bifacial contact configuration. The adopted $J_{0,\ metal}$, $J_{0,\ non\text{-}metal}$, and R_c values are based on own measurements (see text). The estimated current recombination density $J_{0,\ rear}$ and the correspondingly estimated contact resistance $R_{c,\ rear}$ of our developed hole-extracting passivated contact (assuming that the observed screen-printing issues have been solved) are inserted within the iso-efficiency plot (blue dot, full-area contact; black square, bifacial contact). The corresponding practical solar cell efficiency potential using our developed electron- and hole-extracting SiO_x/poly-Si passivated contacts is 22.3%, if a full-area rear-side contact is deployed, and 23.2%, if a bifacial contact is deployed.

electron-selective passivated contact (100 fA cm^{-2}), and we utilized measured $R_{c,\ rear}$ values which we extracted using a thermal evaporated Ag contact instead of a screen-printed contacts, i.e., obtaining $R_{c,\ rear}$ values of 0.225 and 13.5 mΩ cm^2 for the bifacial and full-area rear-contacts, respectively. The corresponding practical efficiency potential, as a function of the quality of the rear-side hole-extracting passivated contact ($J_{0,\ rear}$ and $R_{c,\ rear}$), is shown in **Figure 25**.

Comparing a front-side electron-extracting passivated contact to a conventionally applied front-side hole-extracting diffused contact (front-side boron-diffused emitter, passivated with ALO_x/SiN_x and metallized by bifacial screen printing) greatly improves the front surface passivation quality, reducing the $J_{0,\ front}$ value from ~131 to ~12 fA cm^{-2}, respectively. This can be (1) attributed to the excellent passivation quality of the developed electron-selective passivated contacts itself (6.65 fA cm^{-2} on a textured silicon surface), which cannot be attained by conventional boron diffusion and AlO_x/SiN_x capping (~45 fA cm^{-2}). Furthermore, the metal front-side contacts are now passivated (assuming $J_{0,\ metal}$, ~100 fA cm^{-2}, after industrial screen printing, as measured on 250-nm-thick poly-Si(n^+) capping layers) instead of directly touching the doped silicon wafer ($J_{0,\ metal}$, ~1480 fA cm^{-2}). Therefore, a double-sided passivated contact solar cell has a good potential to obtain higher V_{OC} values at the cell level than a rear-side-only passivated contact solar cell. To give an example, a bifacial double-sided passivated contact solar cell exhibits much lower total surface recombination ($J_{0,\ front} + J_{0,\ rear}$ values of 32.7 fA cm^{-2}) than a bifacial rear-side-only passivated contact solar cell ($J_{0,\ front} + J_{0,\ rear}$ values of ~139.5 fA cm^{-2}), which is ~4 times lower. The improved surface passivation should also directly translate to higher cell efficiencies, which is clearly shown comparing **Figure 25** to **Figure 24**. According to **Figure 25**, a double-sided passivated contact solar cell, using our developed SiO_x/poly-Si(n$^+$) and SiO_x/poly-Si(p$^+$) passivated contacts, exhibits a practical solar cell efficiency potential of ~22.3 and ~23.2%, respectively, using full-area or bifacial rear-side contacts. To recap, the corresponding practical efficiency potential in case of a rear-side-only passivated contact solar cell was 22.3 and 22.5%, respectively. Thus, in case of adopting a bifacial metallization scheme, a double-sided passivated contact solar cell is able to clearly outperform a rear-side-only passivated contact solar cell (practical efficiency potential of ~23.2% as compared to 22.5%, using our developed contact passivation layers).

Again, the bifacial contact scheme appears more advantageous than deploying full-area rear-side contacts, exhibiting a significant 0.9% absolute (4% relative) increase in practical efficiency potential (analyzing double-sided passivated contact solar cells). If we compare bifacial silicon solar cells with a double-sided passivated contact scheme to rear-side-only passivated contact scheme, a respectable gain in cell efficiency by 0.7% absolute (~3% relative) is attainable. Interestingly, if we compare silicon solar cells which utilized full-area rear-side metal contacts, the practical cell efficiency potential for the double-sided passivated contact cell appears to be comparable to the rear-side-only passivated contact cell (both efficiency potentials are in the range of 22.3%). Given comparable $J_{0,\ rear}$, $R_{c,\ rear}$, and $R_{c,\ front}$ values between the two schemes, it seems to indicate that a solar cell adopting a full-area rear-side passivated contact scheme exhibits a low sensitivity of the $J_{0,\ front}$ values on the potential cell efficiency over a range of 12–131 fA cm^{-2} (i.e., the full-area rear-side contact is then limiting the cell efficiency). However, when bifacial contacts are considered, the performance gain by applying additional front-side passivation is substantial. This can be mainly attributed to the reduced recombination underneath the front-side solar cell contacts (further suppressing front-side recombination from 131 to 12 fA cm^{-2} while maintaining a low front-side contact resistance, i.e., comparing 5–13 mΩ cm^2).

One suggestion to further improve the cell efficiency is to utilize laser-assisted local openings into the rear-side dielectrics (as demonstrated in **Figure 22**) and apply a full-area non-fire-through metal contact, which is expected to improve the rear interface reflectance and the corresponding collectable photocurrents.

To summarize, the net surface passivation quality on both the solar cell front-side and rear-side can be significantly improved by incorporating our in-house developed carrier-selective passivated contacts. A double-sided passivated contact scheme is predicted to deliver a ~3% relative improvement of solar cell performance, as compared to a rear-side-only passivated contact scheme. Using a rear-side-only passivated contact scheme, i.e., deploying our in-house developed SiO_x/poly-Si(n$^+$) passivated contact layers and applying conventional bifacial screen printing, we have realized a solar cell efficiency of 21.7% (exhibiting a practical efficiency potential of 22.5%, using our standard boron-diffused front-side contact). The still prevailing challenge is to realize an industrial feasible metallization scheme on hole-extracting poly-Si(p$^+$) contact passivation layers, i.e., to develop suitable pastes to contact p-doped poly-Si by means of screen printing.

4. Conclusion

In this work, we demonstrate the potential of incorporating our in-house developed industrial relevant electron-selective (thermal-SiO_x/poly-Si(n^+)) and hole-selective (thermal-SiO_x/poly-Si(p^+)) passivated contacts into double-sided passivated contact solar cells. Using measured properties of our developed contact passivation layers (i.e., determining the recombination current density j_0 and the contact resistance R_c), we predict a practical efficiency potential approaching 24%, if device integrating them into a front-side textured, electron-extracting, and rear-side planar, hole-extracting solar cell architecture, applying conventional screen printing for contact formation (using a n-type 6-inch Cz wafer with a resistivity of 3.4 Ω cm). Thus far, we have reached a solar cell efficiency of 21.7%, rear side only integrating an electron-extracting SiO_x/poly-Si(n^+) passivated contact and using conventional screen printing.

Our methodology of developing/optimizing (ultrathin) contact passivation layers is outlined as follows: First, we were comparing different tunnel oxides for their suitability to form passivated contacts when capped with highly doped poly-Si, i.e., we analyzed ultrathin (<1.5 nm) industrial relevant SiO_x tunnel layers (i.e., wet-chemically formed silicon oxide (wet-SiO_x), UV/ozone photo-oxidation-formed silicon oxides (ozone-SiO_x), and in situ formed thermal silicon oxides, using low-pressure chemical vapor deposition (LPCVD) (thermal-SiO_x)). Combining specifically designed lifetime and dark I–V test structures, we were able to extract the single-sided saturation current density j_0 and its associated contact resistance R_c for our developed electron-selective and hole-selective passivating contacts. A subsequent optimization of the LPCVD of intrinsic poly-Si capping layers followed by conventional tube diffusion was undertaken, maximizing doping efficiency while minimizing in-diffusion of dopants from the poly-Si capping layer through the SiO_x tunnel layer (which can act as a diffusion barrier) into the silicon wafer. After a subsequent standard SiN_x passivation step, we reached implied open-circuit voltage iV_{oc} values exceeding 730 mV for electron-selective SiO_x/poly-Si(n^+) passivated contacts and exceeding 710 mV for hole-selective SiO_x/poly-Si(p^+) passivated contacts, formed on a planar silicon surface, using an in situ LPCVD-grown thermal-SiO_x tunnel layer prior the LPCVD of the (intrinsic) poly-Si capping layer, and a subsequent tube diffusion. Applying these layers on a textured silicon surface, the electron-extracting SiO_x/poly-Si(n^+) passivated contact still performed well ($iV_{oc} > 700$ mV), whereas the hole-extracting SiO_x/poly-Si(p^+) passivated contact showed unsatisfying performance on a textured surface ($iV_{oc} \sim 630$ mV).

Subsequently, an asymmetric, front-side textured electron-extracting, rear-side planar hole-extracting passivated lifetime structure was processed, reaching an iV_{oc} of 713 mV. Two key challenges have been identified when aiming at a double-sided passivated contact device integration of these layers:

(I) *Parasitic absorption*: There is always a significant amount of parasitic absorption within the poly-Si capping layer, which reduces the absorbed photogeneration current within the silicon wafer, and therefore the maximum possible short-circuit current of the solar cell. Thus, the poly-Si capping layers have to be designed to be as thin as technologically possible. This issue is even by far more important, if aiming at an additional front-side (i.e., double-sided) device integration of passivated contact. Numerical simulations (calibrated toward our developed contact passivation layers) indicate that a 10-nm-thin poly-Si layer still leads to a photogeneration (i.e., short-circuit current) loss of ~1 mA/cm^2, if front-side integrated, whereas the photogeneration loss saturates at 0.05 mA/cm^2 for thicknesses lower than 25 nm, if rear-side integrated. A 25-nm-thin, front-side integrated passivated contact will already exhibit a parasitic absorption loss of 5 mA/cm^2, which is no longer suited for device integration (please note that the relation between the parasitic absorption loss and the thickness of the front-side integrated poly-Si contact passivation layer is rather exponential than linear in this case). In order to develop ultrathin (≤ 10 nm) contact passivation layers, two different process methodologies have been developed: (1) using etch-back technology that is starting from 250-nm-thick poly-Si "standard" layers (as described above) and applying a slow silicon etch (SSE) to reduce the thickness in a controlled way. Using this technology, we were able to obtain ultrathin 3–4-nm hole-extracting SiO_x/poly-Si(p^+) passivated contacts, which are basically maintaining the passivation properties of the thick layers (our corresponding 3–4-nm-thin SiO_x/poly-Si(p^+) contact passivation layers reached an implied open-circuit voltage iV_{oc} of ~690 mV on a planar silicon surface). However, etch-back technology for electron-extracting passivated contacts was possible only down to a thickness of 70 nm. (2) Therefore, we re-optimized the diffusion conditions for ultrathin (10 nm) LPCVD of intrinsic poly-Si layers, which were subsequently subjected to phosphorous tube diffusion in order to obtain 10-nm-thin poly-Si(n^+) electron-extracting capping layers suitable for front-side device integration. Our 10-nm SiO_x/poly-Si(n^+)/SiN_x contact passivation layers reached an implied open-circuit voltage iV_{oc} of 720 mV on a textured silicon surface.

(II) *Compatibility with conventional screen printing*: For "thick" (250–150 nm) electron-extracting SiO_x/poly-Si(n^+)/SiN_x contact passivation layers, conventional screen printing creates no issue. Correspondingly, rear-side-only passivated contact solar cells have been processed, reaching a solar cell efficiency of 21.7% (exhibiting a wet-chemically formed SiO_x tunnel layer and a 250-nm poly-Si(n^+) capping layer further passivated by SiN_x at the rear side of the solar cell and exhibiting a conventional boron-diffused, AlO_x/SiN_x passivated standard emitter at the front side of the solar cell, subsequently being metalized by conventional bifacial screen printing).

However, contacting hole-extracting poly-Si(p^+) layers or ultrathin electron-extracting poly-Si(n^+) layers is a challenge. Trying to contact hole-extracting poly-Si(p^+) layers by screen printing, using conventional fire-through Ag/Al pastes (as used to contact p-doped silicon material), we observe several local "punch-through" contact regions, where the paste is completely consuming the underlying poly-Si(p^+) capping layer, causing a severe degradation of contact passivation quality underneath the metal contact (i.e., there is no more contact passivation). This issue can be attributed to local aluminum alloying processes, which take place during fast-firing of Al containing screen-printing pastes: Al alloying is known to partially consume crystalline silicon material; thus, our thin poly-Si(p^+) capping layers will be consumed upon contact firing, leading to the just outlined local "punch-through" effects. Therefore, the chemical composition of the screen-printing paste itself has to be altered, in order to enable a subsequent damage-free contacting of our (thick or ultrathin) poly-Si(p^+) capping layers. Corresponding research activities, in cooperation with a paste manufacturer, are currently initiated.

Furthermore, it seems that our current ultrathin, 10-nm electron-extracting poly-Si(n^+) layers are not firing stable, especially if deploying high peak firing temperatures (they still do outperform conventionally diffused front-side contacts, though). Interestingly, this is not the case for our "standard" 250-nm-thick layers. ECV measurements confirm that after contact firing (fast-firing in order to form low resistivity contacts), the dopants within the poly-Si(n^+) capping layer have out-diffused into the silicon wafer bulk, thereby effectively reducing field-effect passivation and thus the observed implied open-circuit voltage of the samples after contact firing. Thus, more efforts to render our ultrathin contact passivation layers firing stable, i.e., by deploying lower peak firing temperatures and/or changing the chemical composition of the ultrathin LPCVD of poly-Si capping layers itself, are necessary.

An alternative work plan is to investigate low-temperature metallization approaches, like inline plating.

Nevertheless, despite still having to solve a suited industrial metallization scheme for our ultrathin (≤ 10-nm) in-house developed industrial electron- and hole-selective SiO_x/poly-Si/SiN_x passivated contact layers, due to their excellent passivation and contact resistance properties, these layers have a huge potential to get device integrated into a double-sided passivated contact solar cell architecture, which exhibits a practical efficiency potential of 23.2%, using our measured layer properties for a corresponding numerical prediction. Double-sided passivated contact solar cells deploying bifacial contacts are definitely able to outperform rear-side-only passivated contact solar cells in the near future.

Acknowledgements

This research work was supported by the EIRP-07 project "Passivated contacts for high-efficiency silicon wafer based solar cells" NRF2014EWT-EIRP001-006, in a joint collaboration between SERIS and REC Solar Pte. Ltd.

Author details

Zhi Peng Ling*, Zheng Xin, Puqun Wang, Ranjani Sridharan, Cangming Ke and Rolf Stangl
Solar Energy Research Institute of Singapore (SERIS), National University of Singapore (NUS), Singapore

*Address all correspondence to: lingzhipeng@yahoo.com

References

[1] Fischer M. International Technology Roadmap for Photovoltaic (ITRPV). 2018. Available from: http://www.itrpv.net/

[2] Inns D. Understanding metal induced recombination losses in silicon solar cells with screen printed silver contacts. Energy Procedia. 2016;**98**:23-29

[3] Edler A, Mihailetchi VD, Koduvelikulathu LJ, Comparotto C, Kopecek R, Harney R. Metallization–induced recombination losses of bifacial silicon solar cells. Progress in Photovoltaics. 2015;**23**:620-627

[4] Melskens J, van de Loo BWH, Macco B, Black LE, Smit S, Kessels WMM. Passivating contacts for crystalline silicon solar cells: From concepts and materials to prospects. IEEE Journal of Photovoltaics. 2018;**8**:373-388

[5] Ling ZP, Ge J, Stangl R, Aberle AG, Mueller T. Detailed micro Raman spectroscopy analysis of doped silicon thin film layers and its feasibility for heterojunction silicon wafer solar cells. Journal of Materials Science and Chemical Engineering. 2013;**1**:1-14

[6] Ling ZP, Duttagupta S, Ma F, Mueller T, Aberle AG, Stangl R. Three-dimensional numerical analysis of hybrid heterojunction silicon wafer solar cells with heterojunction rear point contacts. AIP Advances. 2015;**5**:077124

[7] Ling ZP, Mueller T, Aberle AG, Stangl R. Development of a conductive distributed bragg reflector for heterojunction solar cells using n-doped microcrystalline silicon and aluminum-doped zinc oxide films. IEEE Journal of Photovoltaics. 2014;**4**:1320-1325

[8] Tang CG, Ang MCY, Choo K-K, Keerthi V, Tan J-K, Syafiqah MN, et al. Doped polymer semiconductors with ultrahigh and ultralow work functions for ohmic contacts. Nature. 2016;**539**: 536-540

[9] Yoshikawa K, Kawasaki H, Yoshida W, Irie T, Konishi K, Nakano K, et al. Silicon heterojunction solar cell with interdigitated back contacts for a photoconversion efficiency over 26%. Nature Energy. 2017;**2**:17032

[10] Green MA, Emery K, Hishikawa Y, Warta W, Dunlop ED. Solar cell efficiency tables (version 47). Progress in Photovoltaics. 2016;**24**:3-11

[11] F. Haase and R. Peibst. 26.1% record efficiency for p-type crystalline Si solar cells. 2018. Available from: https://isfh.de/en/26-1-record-efficiency-for-p-type-crystalline-si-solar-cells/

[12] Richter A, Benick J, Feldmann F, Fell A, Hermle M, Glunz SW. n-Type Si solar cells with passivating electron contact: Identifying sources for efficiency limitations by wafer thickness and resistivity variation. Solar Energy Materials and Solar Cells. 2017;**173**: 96-105

[13] Zielke D, Petermann JH, Werner F, Veith B, Brendel R, Schmidt J. Contact passivation in silicon solar cells using atomic-layer-deposited aluminum oxide layers. Physica Status Solidi—R. 2011;5: 298-300

[14] Tao Y, Upadhyaya V, Chen C-W, Payne A, Chang EL, Upadhyaya A, et al. Large area tunnel oxide passivated rear contact n-type Si solar cells with 21.2% efficiency. Progress in Photovoltaics: Research and Applications. 2016;**24**: 830-835

[15] Loozen X, Larsen JB, Dross F, Aleman M, Bearda T, O'Sullivan BJ, et al. Passivation of a Metal Contact with a Tunneling Layer. Energy Procedia. 2012;**21**:75-83

[16] Feldmann F, Bivour M, Reichel C, Hermle M, Glunz SW. Passivated rear contacts for high-efficiency n-type Si solar cells providing high interface passivation quality and excellent transport characteristics. Solar Energy Materials and Solar Cells. 2014;**120** (Part A):270-274

[17] Moldovan A, Feldmann F, Zimmer M, Rentsch J, Benick J, Hermle M. Tunnel oxide passivated carrier-selective contacts based on ultra-thin SiO_2 layers. Solar Energy Materials and Solar Cells. 2015;**142**:123-127

[18] Glunz SW, Feldmann F, Richter A, Bivour M, Reichel C, Steinkemper H, et al. The irresistible charm of a simple current flow pattern −25% with a solar cell featuring a full-area back contact. In: Presented at the 31st European Photovoltaic Solar Energy Conference and Exhibition, Hamburg, Germany. 2015

[19] Peibst R, Römer U, Larionova Y, Rienäcker M, Merkle A, Folchert N, et al. Working principle of carrier selective poly-Si/c-Si junctions: Is tunnelling the whole story? Solar Energy Materials and Solar Cells. 2016;**158**: 60-67

[20] Yan D, Cuevas A, Bullock J, Wan Y, Samundsett C. Phosphorus-diffused polysilicon contacts for solar cells. Solar Energy Materials and Solar Cells. 2015; **142**:75-82

[21] Duttagupta S, Nandakumar N, Padhamnath P, Buatis JK, Stangl R, Aberle AG. monoPoly™ cells: Large-area crystalline silicon solar cells with fire-through screen printed contact to doped polysilicon surfaces. Solar Energy Materials and Solar Cells. 2018;**187**:76-81

[22] Reichel C, Müller R, Feldmann F, Richter A, Hermle M, Glunz SW. Influence of the transition region between p- and n-type polycrystalline silicon passivating contacts on the performance of interdigitated back contact silicon solar cells. Journal of Applied Physics. 2017;**122**:184502

[23] Stodolny MK, Anker J, Geerligs BLJ, Janssen GJM, van de Loo BWH, Melskens J, et al. Material properties of LPCVD processed n-type polysilicon passivating contacts and its application in PERPoly industrial bifacial solar cells. Energy Procedia. 2017;**124**:635-642

[24] Peibst YLR, Reiter S, Orlowski N, Schäfer S, Turcu M, Min B, et al. Industrial, screen-printed double-side contacted polo cells. In: 33rd European Photovoltaic Solar Energy Conference and Exhibition, Amsterdam, Netherlands. 2017. pp. 451-454

[25] Yang Y, Altermatt PP, Cui Y, Hu Y, Chen D, Chen L, et al. Effect of carrier-induced hydrogenation on the passivation of the poly-Si/SiO_x/c-Si interface. AIP Conference Proceedings. 2018;**1999**:040026

[26] Yang G, Ingenito A, Hameren NV, Isabella O, Zeman M. Design and application of ion-implanted polySi passivating contacts for interdigitated back contact c-Si solar cells. Applied Physics Letters. 2016;**108**:033903

[27] Green MA, Hishikawa Y, Dunlop ED, Levi DH, Hohl-Ebinger J, Ho-Baillie AWY. Solar cell efficiency tables (version 52). Progress in Photovoltaics: Research and Applications. 2018;**26**: 427-436

[28] Ling ZP, Xin Z, Ke C, Buatis KJ, Duttagupta S, Lee JS, et al. Comparison and characterization of different tunnel layers, suitable for passivated contact formation. Japanese Journal of Applied Physics. 2017;**56**:08MA01

[29] Xin Z, Ling ZP, Nandakumar N, Kaur G, Ke C, Liao B, et al. Surface passivation investigation on ultra-thin

atomic layer deposited aluminum oxide layers for their potential application to form tunnel layer passivated contacts. Japanese Journal of Applied Physics. 2017;**56**:08MB14

[30] Macco B, Loo BWH v d, Kessels WMM. Atomic layer deposition for high-efficiency crystalline silicon solar cells. In: Bachmann J, editor. Atomic Layer Deposition in Energy Conversion Applications. Weinheim: Wiley-VCH; 2017. pp. 41-99

[31] Feldmann F, Ritzau K-U, Bivour M, Moldovan A, Modi S, Temmler J, et al. High and low work function materials for passivated contacts. Energy Procedia. 2015;**77**:263-270

[32] Ke C, Xin Z, Ling ZP, Aberle AG, Stangl R. Numerical investigation of metal-semiconductor-insulator-semiconductor passivated hole contacts based on atomic layer deposited AlO x. Japanese Journal of Applied Physics. 2017;**56**:08MB08

[33] Kaur G, Dwivedi N, Zheng X, Liao B, Ling ZP, Danner A, et al. Understanding surface treatment and ALD AlO_x thickness induced surface passivation quality of c-Si Cz wafers. IEEE Journal of Photovoltaics. 2017;**7**:1224-1235

[34] Xin Z, Ling ZP, Wang P, Ge J, Ke C, Choi KB, et al. Ultra-thin atomic layer deposited aluminium oxide tunnel layer passivated hole-selective contacts for silicon solar cells. Solar Energy Materials and Solar Cells. 2019;**191**:164-174

[35] Werner F, Schmidt J. Manipulating the negative fixed charge density at the c-Si/Al_2O_3 interface. Applied Physics Letters. 2014;**104**:091604

[36] Veith B, Dullweber T, Siebert M, Kranz C, Werner F, Harder NP, et al. Comparison of ICP-AlO_x and ALD-Al_2O_3 layers for the rear surface passivation of c-Si solar cells. Energy Procedia. 2012;**27**:379-384

[37] Hoex B, Schmidt J, van de Sanden MCM, Kessels WMM. Crystalline silicon surface passivation by the negative-charge-dielectric Al_2O_3. In: 33rd IEEE Photovoltaic Specialists Conference. 2008. pp. 1-4

[38] Kotipalli R, Delamare R, Poncelet O, Tang X, Francis LA, Flandre D. Passivation effects of atomic-layer-deposited aluminum oxide. EPJ Photovoltaics. 2013;**4**:45107

[39] Schmidt J, Merkle A, Brendel R, Hoex B, van de Sanden MCM, Kessels WMM. Surface passivation of high-efficiency silicon solar cells by atomic-layer-deposited Al_2O_3. Progress in Photovoltaics. 2008;**16**:461-466

[40] Battaglia C, de Nicolás SM, Wolf SD, Yin X, Zheng M, Ballif C, et al. Silicon heterojunction solar cell with passivated hole selective MoO_x contact. Applied Physics Letters. 2014;**104**:113902

[41] Battaglia C, Yin X, Zheng M, Sharp ID, Chen T, McDonnell S, et al. Hole selective MoO_x contact for silicon solar cells. Nano Letters. 2014;**14**:967-971

[42] Woojun Y, James EM, Eunhwan C, David S, Nicole AK, Erin C, et al. Hole-selective molybdenum oxide as a full-area rear contact to crystalline p-type Si solar cells. Japanese Journal of Applied Physics. 2017;**56**:08MB18

[43] Bivour M, Temmler J, Steinkemper H, Hermle M. Molybdenum and tungsten oxide: High work function wide band gap contact materials for hole selective contacts of silicon solar cells. Solar Energy Materials and Solar Cells. 2015;**142**:34-41

[44] Gerling LG, Mahato S, Morales-Vilches A, Masmitja G, Ortega P, Voz C,

et al. Transition metal oxides as hole-selective contacts in silicon heterojunctions solar cells. Solar Energy Materials and Solar Cells. 2016;**145**: 109-115

[45] Gerling LG, Masmitja G, Ortega P, Voz C, Alcubilla R, Puigdollers J. Passivating/hole-selective contacts based on V_2O_5/SiO_x stacks deposited at ambient temperature. In: 7th International Conference on Silicon Photovoltaics, Freiburg, Germany. 2017. pp. 584-592

[46] Ravindra P, Mukherjee R, Avasthi S. Hole-selective electron-blocking copper oxide contact for silicon solar cells. IEEE Journal of Photovoltaics. 2017;**7**: 1278-1283

[47] Zielke D, Pazidis A, Werner F, Schmidt J. Organic-silicon heterojunction solar cells on n-type silicon wafers: The BackPEDOT concept. Solar Energy Materials and Solar Cells. 2014;**131**:110-116

[48] He J, Gao P, Ling Z, Ding L, Yang Z, Ye J, et al. High-efficiency silicon/organic heterojunction solar cells with improved junction quality and interface passivation. ACS Nano. 2016;**10**: 11525-11531

[49] Ling ZP, Xin Z, Kaur G, Ke C, Stangl R. Ultra-thin ALD-AlO_x/PEDOT: PSS hole selective passivated contacts: An attractive low cost approach to increase solar cell performance. Solar Energy Materials and Solar Cells. 2018; **185**:477-486

[50] Greiner MT, Lu Z-H. Thin-film metal oxides in organic semiconductor devices: Their electronic structures, work functions and interfaces. NPG Asia Materials. 2013;**5**:e55

[51] Meyer J, Hamwi S, Kröger M, Kowalsky W, Riedl T, Kahn A. Transition metal oxides for organic electronics: Energetics, device physics and applications. Advanced Materials. 2012;**24**:5408-5427

[52] Zielke D, Niehaves C, Lövenich W, Elschner A, Hörteis M, Schmidt J. Organic-silicon solar cells exceeding 20% efficiency. Energy Procedia. 2015; **77**:331-339

[53] PV Lighthouse: SunSolve™ 2018. Available from: https://www.pvlighthouse.com.au

[54] Brendel R, Rienaecker M, Peibst R. A quantitative measure for the carrier selectivity of contacts to solar cells. In: 32nd European Photovoltaic Solar Energy Conference and Exhibition, Munich, Germany. 2016. pp. 447-451

[55] Angermann H, Wolke K, Gottschalk C, Moldovan A, Roczen M, Fittkau J, et al. Electronic interface properties of silicon substrates after ozone based wet-chemical oxidation studied by SPV measurements. Applied Surface Science. 2012;**258**:8387-8396

[56] Vitkavage SC, Irene EA, Massoud HZ. An investigation of Si-SiO_2 interface charges in thermally oxidized (100), (110), (111), and (511) silicon. Journal of Applied Physics. 1990;**68**: 5262-5272

[57] Angermann H. Conditioning of Si-interfaces by wet-chemical oxidation: Electronic interface properties study by surface photovoltage measurements. Applied Surface Science. 2014;**312**:3-16

[58] Aberle AG, Glunz S, Warta W. Impact of illumination level and oxide parameters on Shockley-Read-Hall recombination at the Si-SiO_2 interface. Journal of Applied Physics. 1992;**71**: 4422-4431

[59] Aleksandrov OV, Dus' AI. A model of formation of fixed charge in thermal silicon dioxide. Semiconductors. 2011; **45**:467-473

[60] Angermann H. Passivation of structured p-type silicon interfaces: Effect of surface morphology and wet-chemical pre-treatment. Applied Surface Science. 2008;**254**:8067-8074

[61] Gaspard F, Halimaoui A, Sarrabayrouse G. Electrical properties of thin anodic silicon dioxide layers grown in pure water. Revue de Physique Appliquée. 1987;**22**:65-69

[62] Grant NE, McIntosh KR. Passivation of a (100) Silicon Surface by Silicon Dioxide Grown in Nitric Acid. IEEE Electron Device Letters. 2009;**30**: 922-924

[63] Grant NE, McIntosh KR. Silicon surface passivation by anodic oxidation annealed at 400°C. ECS Journal of Solid State Science and Technology. 2014;**3**: P13-P16

[64] Imamura K, Takahashi M, Asuha Y, Hirayama SI, Kobayashi H. Nitric acid oxidation of Si method at 120°C: HNO_3 concentration dependence. Journal of Applied Physics. 2010;**107**:054503

[65] Stodolny MK, Lenes M, Wu Y, Janssen GJM, Romijn IG, Luchies JRM, et al. n-Type polysilicon passivating contact for industrial bifacial n-type solar cells. Solar Energy Materials and Solar Cells. 2016;**158**:24-28

[66] Voutsas AT, Hatalis MK, Boyce J, Chiang A. Raman spectroscopy of amorphous and microcrystalline silicon films deposited by low-pressure chemical vapor deposition. Journal of Applied Physics. 1995;**78**:6999-7006

[67] Pi XD, Gresback R, Liptak RW, Campbell SA, Kortshagen U. Doping efficiency, dopant location, and oxidation of Si nanocrystals. Applied Physics Letters. 2008;**92**:123102

[68] Willoughby AFW, Evans AGR, Champ P, Yallup KJ, Godfrey DJ, Dowsett MG. Diffusion of boron in heavily doped n- and p-type silicon. Journal of Applied Physics. 1986;**59**: 2392-2397

[69] F. Feldmann, S. Mack, B. Steinhauser, L. Tutsch, J.-I. Polzin, J. Temmler, et al. Towards industrial manufacturing of TOPCon 2018. Available from: https://www.pv-tech.org/technical-papers/towards-industrial-manufacturing-of-topcon

[70] Bonilla RS, Hoex B, Hamer P, Wilshaw PR. Dielectric surface passivation for silicon solar cells: A review. Physica Status Solidi (A). 2017; **214**:1700293

[71] McIntosh KR, Johnson LP. Recombination at textured silicon surfaces passivated with silicon dioxide. Journal of Applied Physics. 2009;**105**: 124520-1 to 124520-10

[72] Baker-Finch SC, McIntosh KR. The contribution of planes, vertices, and edges to recombination at pyramidally textured surfaces. IEEE Journal of Photovoltaics. 2011;**1**:59-65

[73] Martin AG. Self-consistent optical parameters of intrinsic silicon at 300 K including temperature coefficients. Solar Energy Materials and Solar Cells. 2008;**92**:1305-1310

[74] Li H, Kim K, Hallam B, Hoex B, Wenham S, Abbott M. $POCl_3$ diffusion for industrial Si solar cell emitter formation. Frontiers in Energy. 2017;**11**:42-51

[75] Deckers J, Loozen X, Posthuma N, O'Sullivan B, Debucquoy M, Singh S, et al. Injection dependent emitter saturation current density measurement under metallized areas using photoconductance decay. In: 28th European Photovoltaic Solar Energy Conference and Exhibition, Paris, France. 2013. pp. 806-810

[76] Hannebauer H, Sommerfeld M, Müller J, Dullweber T, Brendel R.

Analysis of the emitter saturation current density of industrial type silver screen-printed front contacts. In: 27th European Photovoltaic Solar Energy Conference and Exhibition, Frankfurt, Germany. 2012. pp. 1360-1363

[77] SERIS. Griddler 2.5. 2018. Available from: http://www.seris.nus.edu.sg/activities/griddler-2.5.html

Chapter 7

Controllable Synthesis of Few-Layer Graphene on β-SiC(001)

Olga V. Molodtsova, Alexander N. Chaika and Victor Yu. Aristov

Abstract

Few-layer graphene exhibits exceptional properties that are of interest for fundamental research and technological applications. Nanostructured graphene with self-aligned domain boundaries and ripples is one of very promising materials because the boundaries can reflect electrons in a wide range of energies and host spin-polarized electronic states. In this chapter, we discuss the ultra-high vacuum synthesis of few-layer graphene on the technologically relevant semiconducting β-SiC/Si(001) wafers. Recent experimental results demonstrate the possibility of controlling the preferential domain boundary direction and the number of graphene layers in the few-layer graphene synthesized on the β-SiC/Si(001) substrates. Both these goals can be achieved utilizing vicinal silicon wafers with small miscuts from the (001) plane. This development may lead to fabricating new tunable electronic nanostructures made from graphene on β-SiC, opening up opportunities for new applications.

Keywords: silicon carbide, graphene, synthesis, nanodomains, ARPES, LEEM, μ-LEED, XPS, STM

1. Introduction

Extensive studies of mono- and few-layer graphene films, conducted during the last two decades, have revealed unique electronic properties of these low-dimensional materials [1–7], which make them very promising for developing new nanoscale carbon-based electronic technologies [8–13]. Its unique transport properties make graphene a very attractive alternative to silicon in the traditional electronic technologies. However, for successful applications, it is necessary to develop methods of synthesizing low-cost, high-quality graphene films on insulating or semiconducting substrates of sufficiently large size.

Many methods of fabricating ultrathin graphene films have been reported. For example, graphene can be prepared using mechanical or chemical exfoliation from bulk graphite crystals [1, 3, 4, 14]. The mechanically exfoliated graphene layers demonstrate exceptional properties of two-dimensional electron gas, such as extremely high mobility of the charge carriers [15, 16]. However, the exfoliated graphene layers are hardly suitable for technological purposes. The procedures using unzipping of carbon nanotubes, reduction of graphite oxide, chemical vapor deposition, and high-temperature thermal graphitization of single-crystalline substrates were developed to fabricate large-area graphene films [1, 17–22]. To eliminate possible problems associated with the graphene film transfer from one substrate to another, various methods have been developed for direct growing graphene on the technologically relevant non-conducting substrates [23–30].

The hexagonal silicon carbide (α-SiC) wafers are considered the most promising semiconducting substrates for technological synthesis of high-quality graphene films [31–37]. Ultrathin graphene films are usually fabricated on α-SiC using silicon atom sublimation and graphitization of the carbon-enriched surface layers at temperatures above 1000°C [31]. Epitaxial graphene layers synthesized on α-SiC in ultra-high vacuum (UHV) and argon atmosphere demonstrate 2D electronic properties [38–41], which are nearly equivalent to the properties of ultrathin graphene films mechanically exfoliated from bulk graphite crystals. The angle resolved photoelectron spectroscopy (ARPES) studies of the 11-layer graphene on 6H-SiC(000-1) revealed sharp linear dispersions at the K-points typical of monolayer graphene [42].

However, the high price and small size of the single-crystalline α-SiC wafers are not compatible with commercial applications. In order to reduce the price of SiC wafers, epitaxial growth of cubic silicon carbide (β-SiC) thin films on silicon wafers was proposed in the 1980s [43]. Using this method, β-SiC thin films with thickness of several microns could be grown on standard silicon wafers with diameters above 30 cm [44–47] that is highly appealing for direct integration into existing electronic technologies. Fabrication of ultrathin graphene films on β-SiC surfaces, using high-temperature annealing in UHV, was reported for the first time in 2009, when Miyamoto et al. succeeded in synthesizing few-layer graphene on the β-SiC/Si(011) wafers [48]. Then, a number of works demonstrating the feasibility of graphene synthesis on β-SiC/Si wafers of different orientations have been published [48–103]. Mostly, these studies have been conducted on β-SiC(111) thin films [51–61, 65–81] and single-crystalline SiC(111) wafers [62–64]. However, some studies have been carried out on β-SiC(001) [50, 61, 82–93, 101, 102] and even on polycrystalline β-SiC substrates [94]. Since Si(001) is widely used in electronic devices, few-layer graphene films synthesized on the β-SiC/Si(001) wafers can be fully compatible with the existing lithographic processing technologies.

This chapter is focused on the controllable UHV synthesis of few-layer graphene on the β-SiC thin films grown on the technologically relevant Si(001) wafers. Along with detailed atomic and electronic structure studies we present the recent results which uncover the mechanism of layer-by-layer graphene growth on β-SiC/Si(001) and pave the way to synthesize uniform few-layer graphene nanoribbons with desirable number of layers and self-aligned nanodomain boundaries on the low-cost silicon wafers.

2. Atomic and electronic structure of few-layer graphene synthesized on β-SiC/Si(001)

Few-layer graphene synthesis on the β-SiC/Si(001) wafers was demonstrated for the first time in the near-edge X-ray absorption fine structure (NEXAFS), core-level photoelectron spectroscopy (PES), ARPES, and local scanning tunneling microscopy (STM) experiments [50]. Later, the few-layer graphene formation on the β-SiC/Si(001) substrates during high-temperature annealing in UHV was proved by independent Raman spectroscopy experiments [82]. These works showed quasi-free-standing character of the synthesized graphene overlayers. Raman spectroscopy data also revealed the presence of a large number of defects in the few-layer graphene grown on the β-SiC/Si(001) wafers with the average distance between them on the order of 10 nm [82]. However, the origin of these defects could be uncovered only in comprehensive studies using a set of complementary high-resolution micro-spectroscopic techniques, namely, low energy electron microscopy (LEEM), micro low energy electron diffraction (μ-LEED), core-level PES, ARPES, and atomic-resolution STM [85, 87].

Figure 1 shows typical large-area STM images taken from a β-SiC(001) surface before and after trilayer graphene synthesis in UHV [85, 87]. The image of the β-SiC(001)-c(2 × 2) structure (**Figure 1(a)**) reveals extra carbon atoms (bright protrusions) on the surface and monatomic steps. The root mean square (RMS) roughness analysis of the STM images demonstrates substantial enhancement of the surface roughness after the trilayer graphene synthesis. For comparison, the histograms calculated from the STM images of the same size (100 × 100 nm^2) before and after trilayer graphene synthesis are shown on **Figure 1(d)**, **(i)**, and **(j)**. Note that RMS of

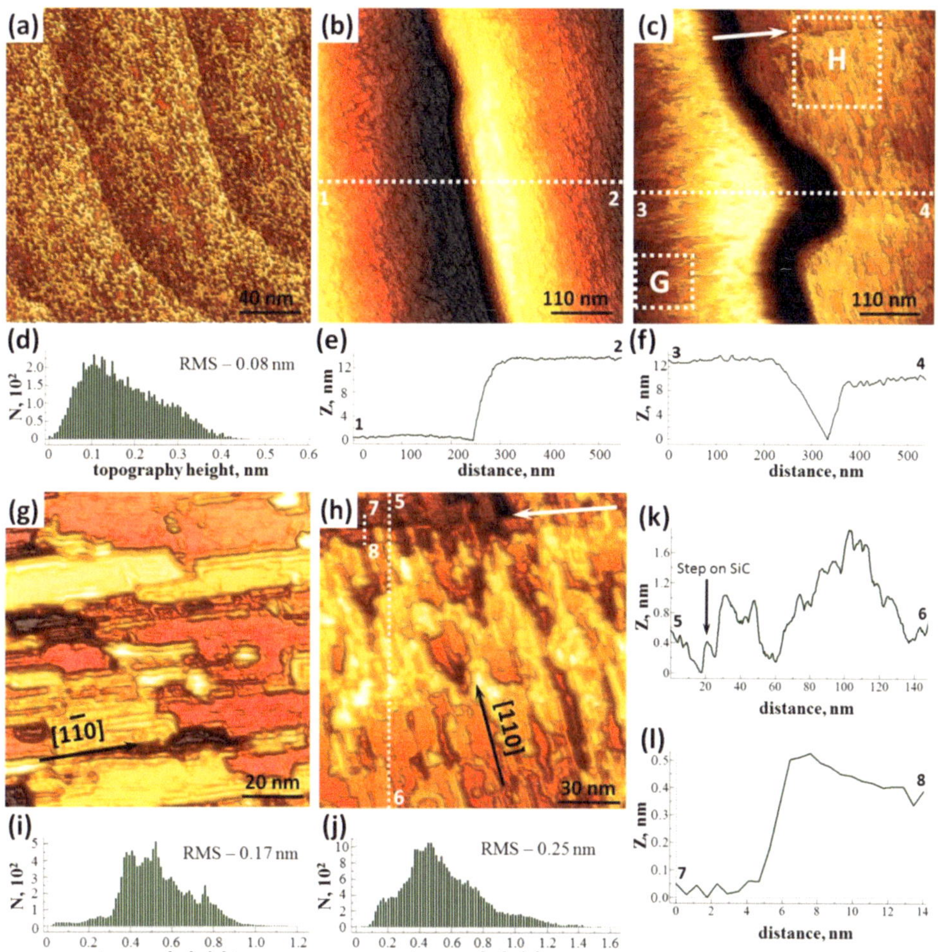

Figure 1.
Large-area STM images of SiC(001)-c(2 × 2) (a) and trilayer graphene/SiC(001) ((b), (c), (g), and (h)). Panels (b) and (c) illustrate the continuity of the graphene overlayer near the multiatomic step (b) and APD boundary (c). The images in panels (g) and (h) emphasize the nanodomains elongated along the [1$\bar{1}$0] (g) and [110] directions (h) observed on the left (area G) and right side (area H) of the APD boundary in panel (c), respectively. The STM images were measured at U = −3.0 V and I = 60 pA (a), U = −1.0 V and I = 60 pA (b), U = −0.8 V and I = 50 pA (c), U = −0.8 V and I = 60 pA (g), and U = −0.7 V and I = 70 pA (h). The white arrows in panels (c) and (h) indicate a monatomic step on the SiC substrate. (d), (i), and (j) Roughness analysis of the STM images in panels (a), (g), and (h). The histograms were calculated from surface areas of the same size (100 × 100 nm^2) for direct comparison of the surface roughness before and after trilayer graphene synthesis. (e), (f), (k), and (l) Cross-sections (1–2), (3–4), (5–6), and (7–8) of the images in panels (b), (c), and (h). Reproduced from Ref. [87] with permission of IOP.

micrometer-scale β-SiC(001)-c(2 × 2) STM images typically varied between 1.0 and 1.5 Å, while the RMS values calculated from the images of trilayer graphene were in the range of 3.0–5.0 Å. The increase of the surface roughness after graphene synthesis is related to the atomic-scale rippling typical for free-standing graphene [104, 105]. STM investigations conducted in different surface areas of several samples [87] confirmed the continuity of the few-layer graphene films covering the β-SiC/Si(001) wafers. As an example, **Figure 1(b)** and **(c)** shows STM images measured at bias voltages corresponding to the bandgap of β-SiC. STM imaging was stable even in the vicinity of multiatomic steps (**Figure 1(b)**) and anti-phase domain (APD) boundaries (**Figure 1(c)**), separating the areas where the β-SiC crystal lattice is rotated by 90°.

STM studies [85, 87] showed that the top graphene layer consists of nanodomains connected to one another through domain boundaries (**Figure 1(g)** and **(h)**). The nanodomain boundaries (NBs) are preferentially aligned with the two orthogonal <110> directions of the SiC crystal lattice, as indicated in **Figure 1(g)** and **(h)**. The domains are elongated in the [110] and [1$\bar{1}$0] directions on the right and left side of the APD boundary, respectively (**Figure 1(c)**). The length and width of the nanodomains on the SiC(001) substrate were in the range of 20–200 nm and 5–30 nm, respectively. These values correlate well with the average distance between defects derived from the Raman spectroscopy studies [82].

The continuity and uniform thickness of the graphene overlayer synthesized on the β-SiC/Si(001) wafers were confirmed in the LEEM experiments (**Figure 2**). The bright-field (BF) LEEM image measured at small electron energy shows uniform contrast throughout the all probed micrometer-sized surface area, including the regions containing defects (steps and APD boundaries). The number of graphene layers can be determined from the number of oscillations in the reflectivity *I-V* curves measured in a small energy window [106, 107]. In the reflectivity spectra shown on **Figure 2(h)**, one can see three reproducible minima, which correspond to the uniform trilayer graphene coverage. The graphene film thickness is homogeneous

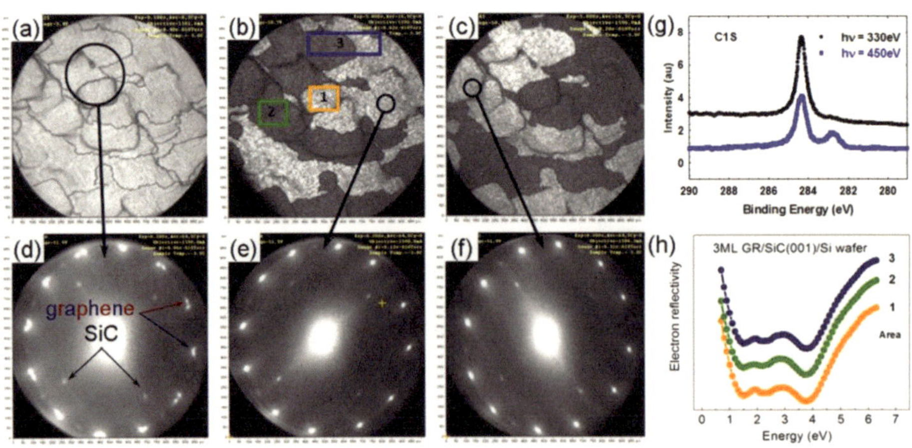

Figure 2.
(a) 20 μm BF LEEM micrograph, recorded with an electron energy of 3.4 eV, proving the uniform thickness of trilayer graphene on β-SiC/Si(001) wafers. (b) and (c) DF LEEM images from different diffraction spots (shown in panels (e) and (f)) demonstrating the contrast reversal on micrometer-scale areas with two rotated graphene domain families. (d)–(f) μ-LEED patterns from the surface areas shown by black circles in panels (a)–(c). The diameters of the sampling areas are 5 μm (a) and 1.5 μm ((b) and (c)), E = 52 eV. (g) μ-PES C 1s spectra taken at two photon energies. The diameter of the probed area is 10 μm. (h) Electron reflectivity spectra recorded for surface regions 1, 2, and 3 as labeled in panel (b). Reproduced from [85] with permission of Tsinghua and Springer.

over the surface and *I-V* curves are almost identical in the surface areas appearing dark and white in the dark-field (DF) LEEM images, as **Figure 2(h)** illustrates.

The μ-LEED patterns taken from micrometer-sized surface areas (**Figure 2(d)**) typically revealed 12 sharp double-split spots and 12 substantially less intense singular spots, corresponding to the nanostructured graphene, and singular spots from the SiC(001) substrate. The μ-LEED patterns measured from different APDs (**Figure 2(e)** and **(f)**) demonstrate 12 non-equidistant spots and six less intense singular graphene spots. The diffraction patterns measured from the APDs are rotated relative to one another by 90°. These LEEM and μ-LEED data demonstrate that each micrometer-sized APD contains graphene nanodomain families with three preferential lattice orientations, giving six preferential lattice orientations on larger (millimeter-scale) surface areas. For the trilayer graphene, four of these six lattice orientations are prevailing in the top layer. The core-level C 1*s* spectra measured from this sample (**Figure 2(g)**) demonstrate only two components with binding energies (BE) corresponding to silicon carbide (lower BE) and graphene (higher BE) in accordance with other core-level PES studies [50], proving the weak interaction of the few-layer graphene with β-SiC(001).

Atomically resolved STM studies presented in **Figure 3** disclose the origin of 12 double-split spots in the LEED patterns (**Figure 2(d)**). STM images in **Figure 3(a)** and **(b)** demonstrate nanodomains elongated in the [110] and $[1\bar{1}0]$ directions, respectively. The 2D fast Fourier transform (FFT) of the STM images consists of two systems of spots, which are related to two graphene lattices rotated by 27°. Inset in **Figure 3(b)** shows one of the FFT patterns. According to the μ-LEED data (**Figure 2(e)** and **(f)**), the graphene domain lattices are

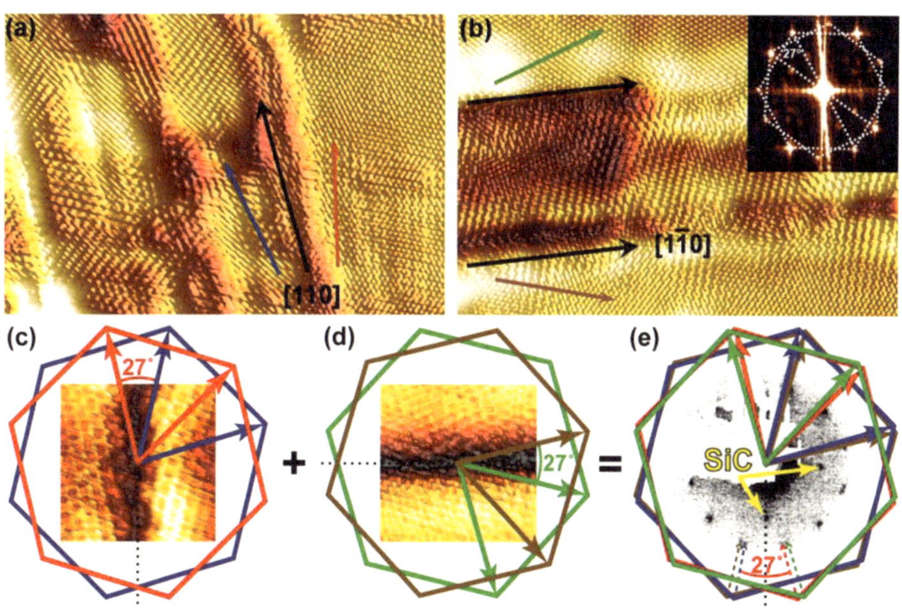

Figure 3.
(a and b) 19.5 × 13 nm² atomically resolved STM images of trilayer graphene nanodomains on SiC(001) elongated along the [110] (a) and $[1\bar{1}0]$ directions (b). The images were taken from different surface areas at U = −10 mV and I = 60 pA. The inset in panel (b) shows an FFT pattern with two 27°-rotated systems of spots. (c–e) Models explaining the origin of the 12 double-split diffraction spots in the LEED pattern shown in Figure 2(d). The insets in panels (c) and (d) are STM images of the <110>-directed domain boundaries. The four differently colored hexagons—red, blue, green, and brown—represent the four preferential domain lattice orientations. The inset in panel (e) shows a LEED pattern taken at E_p = 65 eV, demonstrating 1 × 1 substrate spots (highlighted by yellow arrows) along with 12 double-split graphene spots, indicated by one dotted arrow for each orientation. Reproduced from Ref. [87] with permission of IOP.

preferentially rotated by ±13.5° from the [110] and [1$\bar{1}$0] crystallographic directions of the substrate, which almost coincide with the preferential directions of the nanodomain boundaries. These two families of 27°-rotated domains are rotated by 90° relative to one another and produce two systems of 12 non-equidistant spots in the FFT and μ-LEED patterns (e.g. see **Figure 2(e)** and **(f)**). The sum of two 90°-rotated patterns with 12 non-equidistant spots produce the LEED pattern of graphene/β-SiC/Si(001) with 12 double-split spots, as the models shown in **Figure 3(c–e)** illustrate. These two orthogonal 27°-rotated domain families are usually resolved as horizontal and vertical nanoribbons in STM experiments (**Figure 1(g)** and **(h)**). The DF LEEM images taken from different reflexes in either of the double-split spots show a reversed contrast and confirm that the 27°-rotated domain families typically cover micrometer-sized surface regions in different APDs (**Figure 2(b)** and **(c)**).

Atomic-resolution STM images of the trilayer graphene on β-SiC(001) measured inside the nanodomains usually revealed either hexagonal (**Figure 4(a)**) or honeycomb (**Figure 4(d)**) patterns distorted by atomic-scale rippling [104]. The line profile shown in **Figure 4(c)** reveals random vertical corrugations related to the atomic-scale rippling and regular oscillations with a period of ~ 2.5 Å corresponding to the graphene honeycomb lattice. Typical dimensions of the ripples are about several nanometers laterally and 1 Å vertically (**Figure 4(b)**), coinciding with values predicted by the theory for free-standing monolayer graphene [104]. The random picometer-scale distortions of the sp^2-hybridized carbon bond lengths in graphene/β-SiC/Si(001) are illustrated using smaller area STM images presented in **Figure 4(d–f)**.

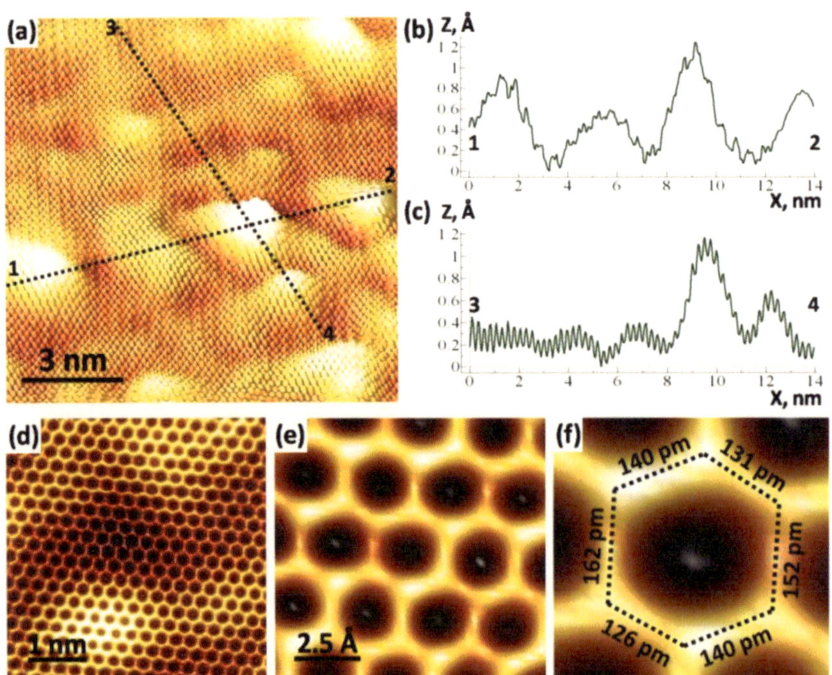

Figure 4.
(a) 13.4 × 13.4 nm² STM image of trilayer graphene on β-SiC(001), illustrating atomic-scale rippling. The image was measured at U = 0.1 V and I = 60 pA. (b) and (c) Cross-sections (1–2) and (3–4) from the image in panel (a). (d–f) STM images of the trilayer graphene, demonstrating random picometer-scale distortions of the honeycomb lattice. The images were measured at U = 22 mV and I = 70 pA (d) and U = 22 mV and I = 65 pA (e and f). One of the distorted hexagons is shown in (f) for clarity. Reproduced from Ref. [103] with permission of Elsevier.

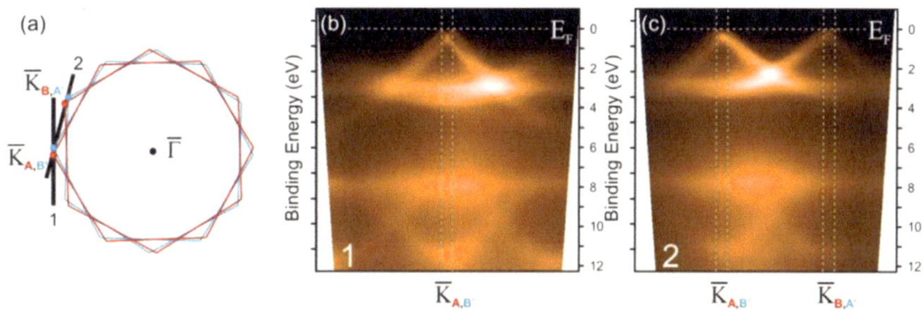

Figure 5.
ARPES characterization of trilayer graphene grown on β-SiC(001). (a) Effective surface Brillouin zone due to superposition of four rotated domain variants (A, B, A′, and B′). (b) and (c) Dispersion of π-band in graphene measured along directions 1 and 2 in (a). Reproduced from Ref. [85] with permission of Tsinghua and Springer.

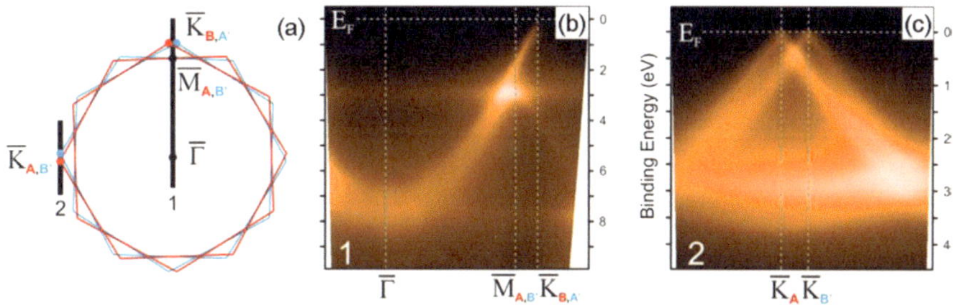

Figure 6.
ARPES characterization of trilayer graphene grown on β-SiC(001). (a) Effective surface Brillouin zone due to superposition of four domain lattices (A, B, A′, and B′). (b) and (c) Dispersion of π-band in trilayer graphene measured by ARPES along directions 1 and 2 in (a). Reproduced from Ref. [85] with permission of Tsinghua and Springer.

The uniformity of the atomic and electronic structure of the trilayer graphene/β-SiC/Si(001) on a millimeter-scale was confirmed by ARPES [85]. As an example, the photoemission studies of the π band are shown in **Figures 5** and **6**. Since ARPES technique probes millimeter-scale sample areas, the effective surface Brillouin zone of graphene on β-SiC(001) comprises Brillouin zones of all rotated lattices (**Figure 5(a)**). The identical sharp linear dispersions typical of quasi-free-standing graphene are observed for all rotated domain variants (**Figure 5(b)** and **(c)**), with the Dirac points located at the Fermi level. The ARPES dispersion measured from the trilayer graphene/β-SiC/Si(001) sample along the $\bar{\Gamma}-\bar{K}$-direction of the surface Brillouin zone (**Figure 6(b)**) reveals the π band reaching the Fermi level. **Figure 6(b)** also displays a dispersion of the π band that backfolds at ~2.5 eV BE and originates from the M-point of the rotated graphene domain. In order to determine the position of the Dirac point, the dispersions were measured in a detection geometry perpendicular to the $\bar{\Gamma}-\bar{K}$-direction (short black line in **Figure 6(a)**) where the interference effects are suppressed and both sides of the Dirac cone are observed [108]. The ARPES data shown in **Figure 6(c)** reveal sharp linear dispersions and tiny additional bands between the two split Dirac cones. According to the theoretical calculations presented in Ref. [93], the observed ARPES dispersions may correspond to a quasi-free-standing Bernal-stacked ABA-trilayer graphene formed on β-SiC(001).

The experimental data presented in this section were obtained during high-resolution studies of the trilayer graphene grown on a low-index β-SiC/Si(001) wafer using high-temperature annealing in UHV. They demonstrate the fabrication of uniform nanostructured graphene with two preferential NB directions on millimeter-sized samples. Such nanoribbon systems supported on the Si(001) wafers are very promising because the presence of the self-aligned boundaries can provide a sizeable energy gap in graphene [10]. However, for technological applications it is highly desirable to control the thickness of the graphene overlayer and reduce the number of the preferential NB orientations from two to one. Note that thickness of the few-layer graphene synthesized on β-SiC/Si(001) wafers by different groups, utilizing very similar UHV thermal treatment procedures, varied from one to several monolayers [50, 82–93, 101].

3. Layer-by-layer graphene growth on β-SiC/Si(001)

For detailed understanding the mechanisms of the surface transformation and layer-by-layer graphene growth on β-SiC/Si(001) in UHV at high temperatures a series of experimental studies with *in-situ* control of the surface atomic and electronic structure during heating have been conducted [102, 109]. **Figure 7** summarizes the β-SiC(001) surface transformations during annealing in UHV, monitored using *in-situ* core-level PES (**Figure 7(a–g)**) and *ex-situ* LEED and STM (**Figure 7(h–l)**).

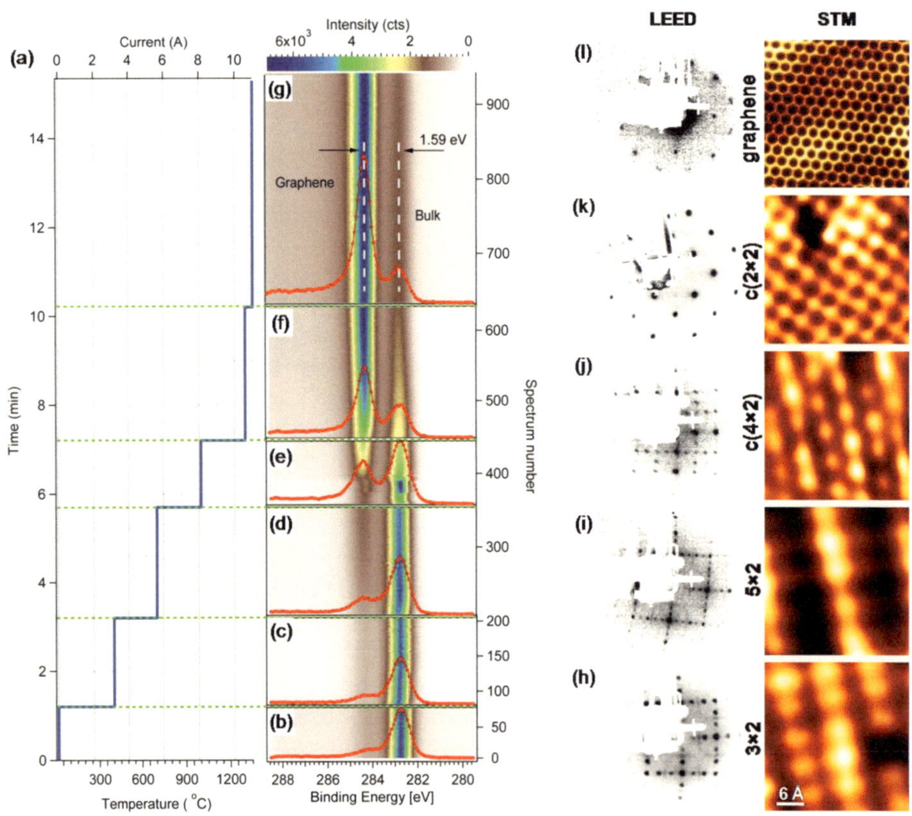

Figure 7.
(a–g) In-situ core-level PES studies of β-SiC/Si(001) during heating in UHV. (a) Temperature of the sample during the PES measurements. (b–g) Time evolution of the C 1s core-level spectra recorded in snapshot regime during heating. A single spectrum taken in the corresponding temperature interval (shown in panel (a)) is presented. (h–l) Evolution of the SiC(001) surface atomic structure probed by LEED and STM. The 3 × 2, 5 × 2, c(4 × 2), and c(2 × 2) reconstructions are consecutively formed on the SiC(001) surface in the temperature range of 800–1300°C before the graphene overlayer formation. Reproduced from Ref. [109] with permission of Elsevier.

The first steps toward successful graphene synthesis on β-SiC/Si(001) relate to the removal of the protective silicon oxide layer and the fabrication of a contaminant-free SiC(001)1 × 1 surface structure. This reconstruction can be fabricated after outgassing the sample holder and flash-heating the β-SiC/Si(001) wafers at 1000–1100°C. Then, the fabrication of a graphene overlayer includes the deposition of several monolayers (MLs) of silicon atoms onto the clean, carbon-rich SiC(001)1 × 1 surface and annealing at gradually increasing temperatures. Depending on the quality of the β-SiC thin film grown on the Si(001) wafer, the deposition and annealing cycles can be repeated until sharp 1 × 1 LEED pattern is observed.

Figure 7(a–g) shows the results of the PES experiments with real-time control during the direct-current sample heating with silicon deposited onto SiC(001)1 × 1 surface structure [109]. During the measurements, a current was applied to heat the sample up to 1350°C (**Figure 7(a)**). The C 1*s* core-level spectra were taken in a snapshot mode during the sample heating with an acquisition time of 1s/spectrum, using a photon energy of 750 eV. Six core-level spectra taken at different stages of the surface graphitization are shown in **Figure 7(b–g)**. Two main C 1*s* peak components can be distinguished in the spectra, which change their relative intensity with increasing temperature. Note that the absolute (but not the relative) binding energies of the individual components in this experiment could be modified by the voltage applied across the β-SiC/Si(001) wafer.

At lower temperatures (**Figure 7(b)**), a strong peak corresponding to the bulk carbon atoms dominates in the PES spectra. At temperatures above 1200°C (**Figure 7(d–f)**), an additional component (shifted to higher BE) starts to grow, while the relative intensity of the bulk component decreases. The change of the C 1*s* core-level shape corresponds to the carbonization of the top surface layers at high temperatures. At temperatures close to the silicon melting point (1350°C), the carbon-carbon bonds undergo a transition to sp^2 hybridization corresponding to graphene lattice formation (**Figure 7(g)**). *Ex-situ* LEED measurements proved the existence of a graphene overlayer on the β-SiC/Si(001) wafer used for the PES experiments presented in **Figure 7(b–g)**.

Figure 7(h–l) shows step-by-step LEED and STM studies of the β-SiC(001) surface atomic structure after heating in UHV at various temperatures. They prove consecutive fabrication of different β-SiC(001) surface reconstructions in accordance with Refs. [110, 111]. The LEED and STM data in **Figure 7(h–l)** were obtained after consecutive heating of the same β-SiC/Si(001) sample in UHV to 1000, 1150, 1200, 1250, and 1350°C, and cooling to room temperature. After annealing at temperatures of 700–1000°C a uniform, Si-rich SiC(001)3 × 2-reconstructed surface with large (001)-oriented terraces is fabricated (**Figure 7(h)**). Increasing the annealing temperature from 1000 to 1250°C leads to consecutive fabrication of the silicon-terminated 5 × 2 (**Figure 7(i)**), c(4 × 2) (**Figure 7(j)**), 2 × 1, and carbon-terminated c(2 × 2) reconstructions (**Figure 7(k)**). According to the LEED and STM studies, the most uniform graphene overlayers on β-SiC(001) can be obtained after flash heating (10–20 s) of the c(2 × 2) reconstruction at 1350°C with post-annealing at 600–700°C, which is similar to the method used for the synthesis of graphene on α-SiC [39, 112, 113]. The LEED pattern shown in **Figure 7(l)** reveals sharp substrate spots and 12 double-split graphene spots related to the formation of the few-layer graphene nanodomain network similar to the one presented in **Figures 1** and **3**.

The exact number of the graphene layers synthesized on β-SiC(001) during UHV heating could strongly depend on the vacuum conditions, annealing temperature and duration [70]. To uncover the mechanism of the layer-by-layer graphene growth on the β-SiC/Si(001) substrates and find the way to control the number of synthesized graphene layers and preferential nanodomain boundary direction, *in-situ* high-resolution core-level and angle-resolved photoelectron spectroscopy, LEEM and μ-LEED studies have been carried out [102].

Figure 8 shows μ-LEED, LEEM *I-V*, ARPES and micro X-ray photoelectron spectroscopy (μ-XPS) data obtained from the same sample region *in-situ* during the high-temperature surface graphitization in UHV. The μ-LEED pattern and C 1s core level spectra taken in bulk- and surface-sensitive regimes from the β-SiC(001)-c(2 × 2) reconstruction prepared prior to graphene synthesis are shown in **Figure 8(a)**. Then, the temperature of the β-SiC/Si(001) wafer was increased and graphene spots were observed in the μ-LEED patterns. The first graphene monolayer (**Figure 8(b)**) was formed after 10 short flashes at temperatures in the range from 1250 to 1300°C and pressures not exceeding 5×10^{-9} mbar. For the fabrication of 2 and 3 ML graphene (**Figure 8(c)** and **(d)**), 50 and 100 flash heating cycles, respectively, were applied at 1300°C. The number of the synthesized graphene layers was defined from the number of minima in the low energy part of the electron reflectivity *I-V* curves presented in **Figure 8** (top). The graphs in **Figure 8** (bottom) depict the evolution of the C 1s spectra acquired in normal emission from a circular sample area (d = 2 μm) at 325, 330, 400, and 450 eV photon energies for a SiC(001)-c(2 × 2) reconstruction **(a)**, mono- **(b)**, bi- **(c)**, and trilayer graphene **(d)**. The selected photon energies correspond to different surface sensitivities of the XPS measurements with the highest sensitivity achieved at 325 and 330 eV. The C 1s spectra were decomposed into individual components corresponding to different carbon atom chemical bonds [102]. The results of the C 1s spectra decomposition are presented in **Figure 8** together with the experimental data (black circles) where the red line is the

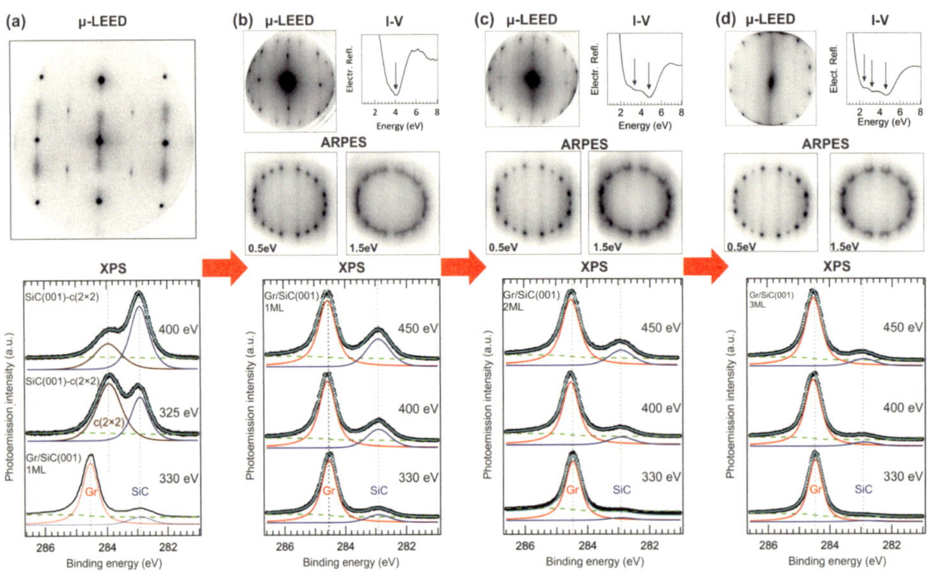

Figure 8.
In-situ LEEM, μ-LEED, ARPES, and XPS showing control over the SiC(001) surface graphitization during heating in UHV. (a) μ-LEED (top) and μ-XPS data (bottom) obtained from the SiC(001)-c(2 × 2) reconstruction. To illustrate the difference between the c(2 × 2) and graphene spectra, the C 1s spectrum from the 1 ML graphene/SiC(001) system is presented (bottom line). (b–d) μ-LEED, LEEM I-V reflectivity spectra, ARPES constant energy maps, and μ-XPS data obtained from one of APDs of the graphene/SiC(001) system at approximately 1 ML (b), 2 ML (c), and 3 ML (d) coverage. Top row: μ-LEED taken from a circular sample area (d = 0.5 μm) using a 44 eV electron beam (left) and LEEM I-V curves demonstrating one (b), two (c), and three (d) minima (indicated by arrows) corresponding to the number of the synthesized graphene layers (right). Middle row: Photoemission angular distribution maps taken at 0.5 eV (left) and 1.5 eV (right) binding energies, measured using a 47 eV photon energy. Bottom row: C 1s core level spectra (black circles) obtained with 330, 400 and 450 eV photon beams for 1, 2 and 3 ML from a circular sample areas (d = 2 μm) and results of the spectra decomposition. Reproduced from Ref. [102] with permission of ACS.

graphene peak (Gr), the blue line is the bulk SiC peak, the brown line is the surface β-SiC(001)-c(2 × 2) component, the green dashed line is the background and the cyan line is the envelope. One can note that each spectrum displays only two main components. The Gr peak is shifted by ~1.65 eV toward higher BE relative to the bulk SiC peak located at 282.9 eV. The intensity of the bulk SiC component decreases both with decreasing photon energy and increasing number of graphene layers. No other components except Gr and SiC were detected in the C 1s spectra, confirming the absence of strong chemical interactions between the graphene overlayer and β-SiC, which would provide additional components with higher BE [103]. The intensities of the individual components in the photoemission spectra shown in **Figure 8** can be used as a reference to distinguish between mono-, bi-, and trilayer graphene on the β-SiC/Si(001) wafers using XPS technique only. Measuring the XPS spectra in the normal emission geometry with 330, 400, and 450 eV photon energies, using the fast dynamic-XPS stations with real-time control of the core level spectra shape [109], one can stop the synthesis procedure when a desirable number of graphene layers (1, 2, or 3 ML) is synthesized.

The *in-situ* ARPES and μ-LEED measurements (**Figure 8**) uncover the origin of the 12 singular spots located between 12 double spots in the LEED pattern taken from the trilayer graphene synthesized on β-SiC(001) (**Figure 2(d)**) and explain the mechanism of the layer-by-layer graphene growth. The singular spots in **Figure 2(d)** are aligned with the SiC substrate spots in contrast with the ±13.5° rotated diffraction patterns corresponding to the rotated graphene nanodomain lattices shown in **Figure 3**. The middle row of images in **Figure 8** shows the ARPES intensity constant-energy maps taken at $E = E_F - 0.5$ eV and $E = E_F - 1.5$ eV as a function of graphene coverage. The ARPES maps prove the conical shape of the Fermi surface for all preferential graphene nanodomain lattice orientations (two non-rotated lattices and four lattices rotated by ±13.5° relative to the two orthogonal <110> directions) at three graphene coverages studied. Notably, both μ-LEED and ARPES maps measured for the 1 ML graphene/SiC(001) system reveal almost the same intensities of the features corresponding to the non-rotated and ± 13.5° rotated domain lattices. The intensity of the diffraction spots and ARPES features corresponding to the non-rotated lattices is systematically suppressed when graphene coverage increases from 1 ML (**Figure 8(b)**) to 2 ML (**Figure 8(c)**), and then to 3 ML (**Figure 8(d)**). The non-rotated graphene lattice orientations are prevailing only at the beginning of the β-SiC(001) surface graphitization. In contrast, when graphene coverage reaches several monolayers, most of the β-SiC(001) surface is covered by nanodomains with four preferential graphene lattice orientations, rotated ±13.5° relative to the two orthogonal <110> directions (**Figure 2(d)**).

Figure 9 shows **(a)** LEEM and **(b–e)** ARPES data obtained from the 1 ML graphene/β-SiC(001) sample. **Figure 9(b)** and **(c)** shows the constant energy ARPES intensity maps measured from different APDs marked as B and C on panel **(a)**. **Figure 9(d)** and **(e)** shows the dispersions obtained by a cut through the experimental data as indicated by the dashed lines in panels **(b)** and **(c)**, respectively. Eighteen identical linear dispersions are clearly resolved, proving that domains with all six preferential lattice orientations at 1 ML graphene coverage exhibit the same electronic structure typical of free-standing monolayer graphene.

The prevalence of the μ-LEED and ARPES features associated with the non-rotated lattices at sub-monolayer coverages is a key to understand the mechanism of the graphene growth on the β-SiC/Si(001) wafers. **Figure 10(a–c)** illustrates how the non-rotated graphene domain lattice can match the SiC(001)-c(2 × 2) reconstruction. If the lattice parameters of the c(2 × 2) square unit cell (red square) are doubled, it matches well to a slightly distorted square (green lines) connecting carbon atoms of the graphene lattice, which can be laterally translated to cover

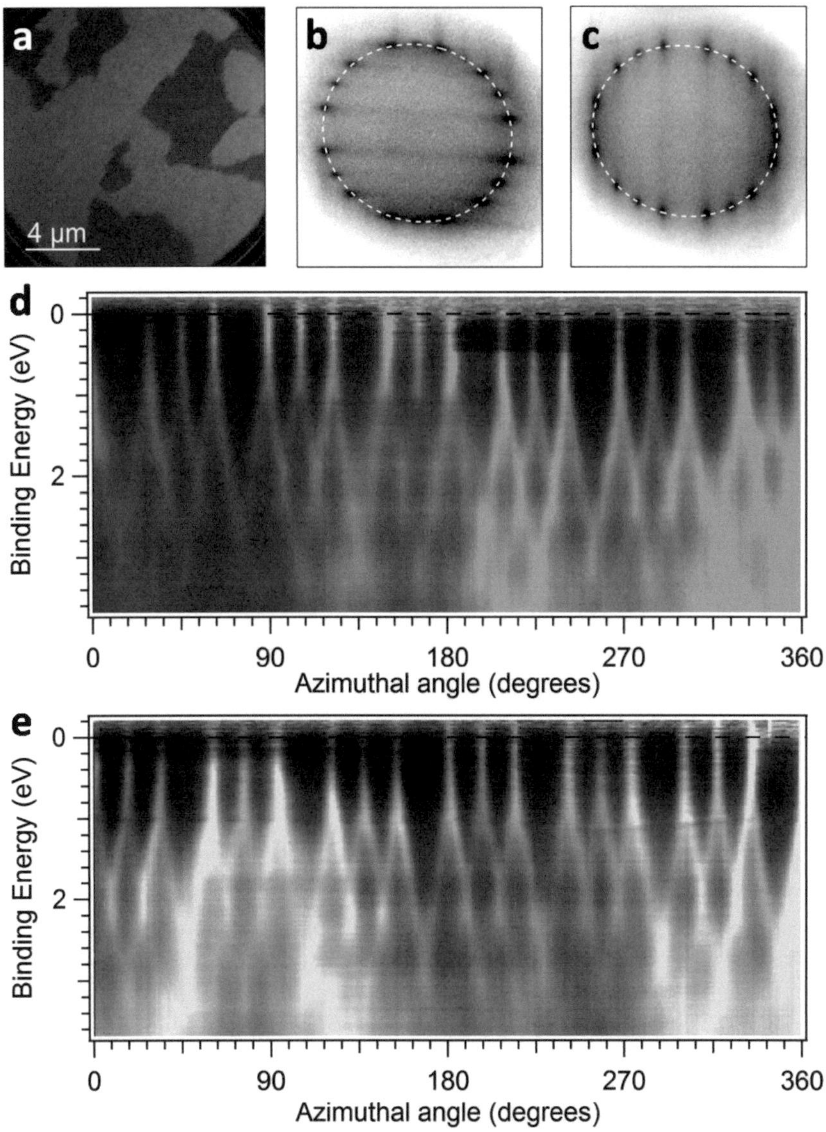

Figure 9.
LEEM and ARPES characterization of the nanostructured monolayer graphene. (a) DF-LEEM taken from the 1ML graphene/SiC(001) system. (b) and (c) Corresponding photoemission patterns taken for domains B and C in panel (a) at $E - E_F$ = 0.5 eV, from a circular sample area (d = 2 μm, hν = 47 eV). (d) and (e) Dispersion of the Dirac cones obtained by a cut through the data as shown by the dashed lines in patterns (b) and (c). Reproduced from Ref. [102] with permission of ACS.

the entire c(2 × 2) surface by the graphene overlayer (**Figure 10(a)** and **(c)**). The mismatch of these two quadrilaterals is below 2%, which is likely sufficient to initiate the growth of the non-rotated graphene monolayer on SiC(001)-c(2 × 2). Such a small mismatch cannot be found for other possible surface structures. Therefore, the SiC(001)-c(2 × 2) reconstruction is a necessary step for successful high-temperature graphene synthesis on β-SiC(001). This is very similar to results of the STM studies performed on β-SiC(111) [71], where the transition from a typical ($\sqrt{3} \times \sqrt{3}$)R30° to an intermediate (3/2 × $\sqrt{3}$)R30° structure matching the graphene (2 × 2) unit cell was observed before the formation of the honeycomb (1 × 1) overlayer.

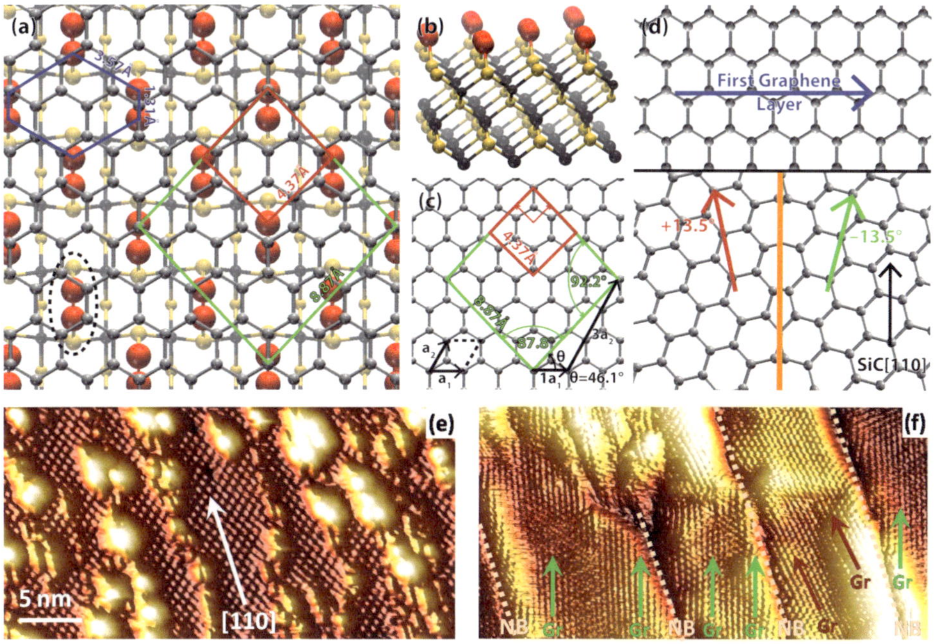

Figure 10.
Model of few-layer graphene growth on β-SiC(001). (a) A schematic model showing the non-rotated graphene lattice on top of the SiC(001)-c(2 × 2) reconstruction. Carbon and silicon atoms are shown as gray and yellow spheres, respectively, with c(2 × 2) carbon dimers highlighted by red spheres. The red square indicates the c(2 × 2) unit cell, the green square shows the distorted coincidence quadrilateral resembling a doubled c(2 × 2) unit cell. (b) Quasi-3D view of the SiC(001)-c(2 × 2) reconstruction. (c) A model of the graphene honeycomb lattice with the quadrilaterals showing the surface and overlayer cells. (d) A schematic model of the few-layer graphene growth on SiC(001): At the beginning, domains with a non-rotated honeycomb lattice nucleate in accordance with panel (a), then ±13.5°-rotated lattices start to grow from the linear defects, which become the nanodomain boundaries in the nanostructured graphene overlayer. (e,f) Atomically resolved STM images of the SiC(001)-c(2 × 2) surface (e) and trilayer graphene synthesized on a β-SiC/Si(001) wafer (f). Reproduced from Ref. [102] with permission of ACS.

The model in **Figure 10** suggests that carbon dimers of the c(2 × 2) reconstruction (indicated by dotted black oval in **Figure 10(a)**) may be considered the smallest building blocks of the non-rotated graphene lattice, since the distance between carbon atoms in the dimers (1.31 Å) is reasonably close to that of the graphene honeycomb lattice (1.46 Å). In order for graphene growth to begin, extra carbon atoms must be present on the c(2 × 2) surface to provide the substantially higher density of carbon atoms in the graphene lattice. Additional carbon atoms are actually observed during the STM studies as random bright protrusions (**Figure 1**) or linear <110>-directed atomic chains decorating the SiC(001)-c(2 × 2) reconstruction (**Figure 10(e)**). These adatoms form chemical bonds with the dimers of the c(2 × 2) reconstruction at high temperatures and initiate the preferential growth of graphene nanodomains with lattices non-rotated relative to the SiC <110> directions. These domains cannot grow to micrometer-scale due to the presence of linear defects on the SiC(001)-c(2 × 2) surface (**Figure 10(e)**) and the mismatch between the c(2 × 2) and graphene lattices producing strain in the overlayer. However, the reasonably small mismatch of the c(2 × 2) and the graphene lattice (**Figure 10(a)**) leads to the prevalence of the two non-rotated lattice variants in the graphene/SiC(001) system until the first monolayer is complete. The next layers presumably grow on top of the first monolayer starting from the linear defects on the surface (either steps or < 110>-directed linear atomic chains), which is supported by the very fast suppression of the non-rotated domain features in the μ-LEED and ARPES maps with increasing graphene coverage (**Figure 8**). The second and third graphene layers can start to grow from the linear defects line-by-line [114] which define the positions and orientations of the nanodomain boundaries in the few-layer graphene/β-SiC(001) (**Figure 10(f)**). In this case, it is energetically favorable for graphene lattices in neighboring nanodomains to be rotated by 27° relative to one another, as the model in **Figure 10(d)** (bottom part) illustrates. The comparison of the atomic resolution STM images of the SiC(001)-c(2 × 2) and trilayer graphene/SiC(001) clearly shows the coincidence of the carbon atomic chain directions in the former structure (**Figure 10(e)**) and nanodomain boundary directions in the latter (**Figure 10(f)**). This result suggests that controlling the density and orientation of defects on β-SiC/Si(001) (e.g., steps on vicinal substrates) could allow the average size of the graphene domains and their orientation to be tuned. This can open a way for synthesis of self-aligned graphene nanoribbons supported by the technologically relevant β-SiC substrate.

4. Fabrication of self-aligned graphene nanoribbons using β-SiC thin films grown on vicinal Si(001) wafers

Synthesis of the uniform self-aligned trilayer graphene nanoribbon structure using β-SiC thin films grown on the vicinal Si(001) wafers with a miscut of 2° was reported in Ref. [93]. STM studies revealed that nanodomains on the vicinal sample are preferentially elongated in one direction (coinciding with the step direction of the bare SiC(001) substrate). This is illustrated in **Figure 11(a)** and **(b)**. Remarkably, the direction of the nanodomain boundaries in the trilayer graphene was the same in different APDs of the 2°-off β-SiC/Si(001) sample [93]. **Figure 11(c)** shows an atomically resolved STM image containing three nanodomains and three boundaries (NB). Detailed analysis of the STM images measured from various graphene/SiC/Si(001) samples showed that NBs are frequently rotated by 3.5° relative to the [110] crystallographic direction as depicted in **Figure 11(e)**. Since the graphene lattices in neighboring nanodomains are rotated by ±13.5° from the same [110] direction, they are asymmetrically rotated relative to the NBs. The lattices in neighboring domains are rotated by 10° counterclockwise (Gr_L)) and 17° clockwise (Gr_R) relative to the

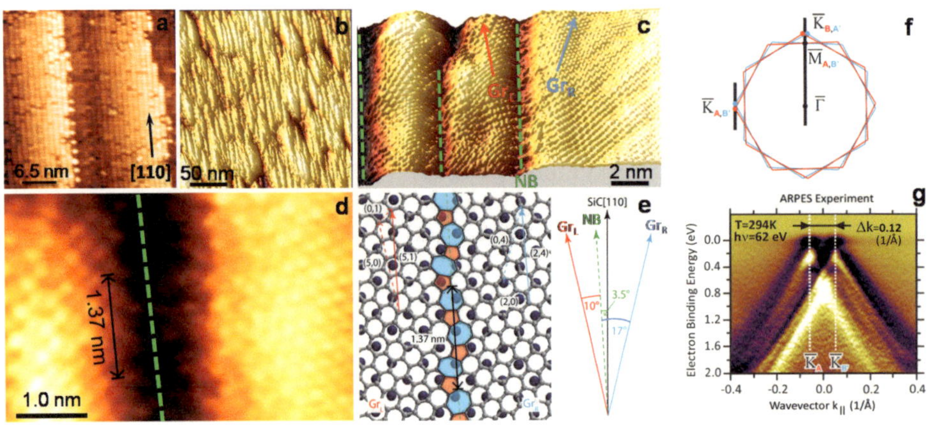

Figure 11.
(a) STM image of the vicinal SiC(001)-3 × 2 surface. The step direction is close to the [110] direction of the SiC crystal lattice. (b) Large-area STM image of graphene nanoribbons synthesized on the vicinal SiC(001). (c) and (d) Atomically resolved STM images of the graphene surface. The domain lattices are rotated 17° clockwise (Gr_R) and 10° anticlockwise (Gr_L) relative to the NB. The NB is itself rotated 3.5° anticlockwise from the [110] direction. (e) Schematic model of the NB for the asymmetrically rotated nanodomains in panels (c) and (d). For the angles shown a periodic structure of distorted pentagons and heptagons is formed. (f) Effective surface Brillouin zone corresponding to four rotated graphene domain variants. (g) Dispersion of the π-band in the graphene along the K_A-K_B direction indicated in panel (f) [93].

NB (**Figure 11(c)**). As **Figure 11(e)** illustrates, this asymmetry leads to the formation of a periodic structure along the boundaries, with a period of 1.37 nm. The periodic structure consists of distorted heptagons and pentagons, which is consistent with the atomically resolved STM image measured at the NB (**Figure 11(d)**). ARPES measurements, conducted on the same sample, showed sharp linear dispersions in the K-points for all preferential graphene lattice orientations (**Figure 11(f)** and **(g)**).

Recent theoretical studies [115] have demonstrated that graphene domain boundaries with a periodic atomic structure can reflect electrons over a large range of energies. This would provide a possibility to control the charge carriers in graphene without the need to introduce an energy bandgap. **Figure 12(a)** shows a schematic of a graphene nanogap device utilized for investigations of the transport properties of graphene synthesized on the β-SiC/2°-off Si(001) wafer [93]. The voltage was applied perpendicular to the nanodomain boundaries to investigate the local transport properties of the self-aligned nanoribbon system with asymmetrically rotated graphene domain lattices (**Figure 11**). According to the theory [115], a charge transport gap of $E_g = \hbar \nu_F \frac{2\pi}{3d} \approx \frac{1.38}{d(\mathrm{nm})}$ (eV) could be induced by a non-symmetric rotation of graphene lattices in neighboring domains, where $\hbar$ is the reduced Planck's constant, ν_F is the Fermi velocity, and d is the periodicity along the NB. As indicated in **Figure 11(d)** and **(e)**, the asymmetric rotation of the graphene lattices in the nanostructured trilayer graphene synthesized on β-SiC/2°-off Si(001) leads to a 1.37 nm periodicity along the NB. The formation of this periodic structure could be responsible for a transport gap of about 1.0 eV, which was observed in the low-temperature transport measurements (**Figure 12**). The transport gap is observed at temperatures below 100 K (**Figure 12(b)** and **(c)**). According to the *dI/dV* spectra shown in **Figure 12(d)**, the transport gap is ~1.3 eV at 50 and 10 K and substantially smaller (~0.4 eV) at 100 K. The conductivity of the trilayer graphene/β-SiC(001) nanogap device is only 0.01 μS at small voltages

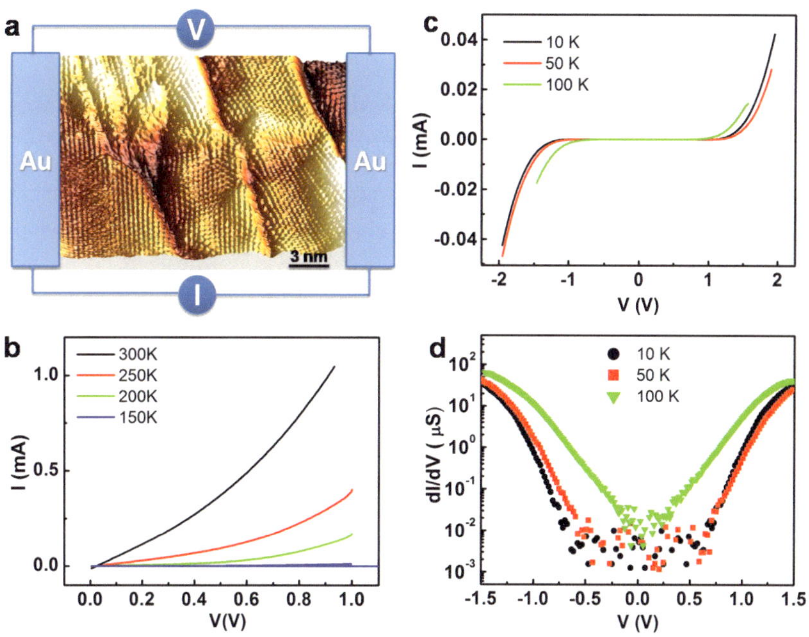

Figure 12.
(a) Schematic drawing of the nanogap device fabricated on the trilayer graphene/β-SiC/2°-off Si(001) sample. (b) I-V curves measured at 150, 200, 250 and 300 K. (c) I-V curves measured at 10, 50, and 100 K. (b) and (c) are measured with the current directed across the self-aligned NBs. (d) Corresponding dI/dV curves for temperatures below 150 K. Reproduced from Ref. [93] with permission of ACS.

and 100 μS at larger voltages when the electric current start to flow. This gives a high on-off current ratio of 10^4.

This successful demonstration of the transport gap opening in the nanostructured trilayer graphene (**Figure 12**) became possible because the NBs were uniformly aligned with the step direction of the vicinal β-SiC(001) substrate (**Figure 11**), i.e., perpendicular to the current direction in the electric measurements. Note that although the NBs with asymmetrical rotation of the graphene lattices were most frequently observed, boundaries with other atomic structures were also resolved in detailed atomic resolution STM studies of several few-layer graphene/β-SiC/Si(001) samples [85, 87, 93, 101, 102]. Despite the differences in the atomic structure, STM images generally revealed extreme distortions of the overlayer near the NBs (e.g., **Figures 1(g)** and **12**). The graphene overlayer in these areas was usually bent upward and downward, forming semi-tubes with typical diameters of several nanometers. According to the STM data, the radii of curvature of the ripples in the few-layer graphene on β-SiC/Si(001) wafers were typically in the range of 2–5 nm [101]. As revealed the theoretical calculations and recent experimental studies [101], the ripples formed at the NBs could also be responsible for the opening transport gap in graphene/β-SiC/Si(001). The self-aligned nanodomain boundaries with ripples can also be utilized to add the spin degree of freedom to graphene [101], since spin-orbit coupling can be induced by the curvature of the ripples [116].

5. Conclusions

We have summarized the results of high-resolution studies of the atomic and electronic structure of few-layer graphene synthesized in UHV on β-SiC thin films epitaxially grown on the technologically relevant Si(001) wafers. LEEM, μ-LEED, ARPES, and STM studies revealed that graphene overlayer on the β-SiC/Si(001) substrates consists of nanodomains with six preferential lattice orientations and two preferential nanodomain boundary directions. The number of the boundary directions can be reduced to one using vicinal wafers with small miscuts from the Si(001) plane. Thus, self-aligned graphene nanoribbon system supported by a wide-gap semiconductor substrate could be fabricated using 2°-off Si(001) wafer. *In-situ* studies of the few-layer graphene synthesis on the β-SiC/Si(001) wafers performed in UHV using micro-spectroscopic methods demonstrate that thickness of the graphene overlayer can be controlled in the course of the high-temperature synthesis and the procedure can be stopped when a desirable number of graphene layers (e.g., 1, 2, or 3 ML) is synthesized.

Acknowledgements

This work was carried out within the state task of ISSP RAS and supported by the Russian Foundation for Basic Research (Grant nos. 17-02-01139, 17-02-01291) and Erasmus plus mobility grant (2016-1-IE02-KA107-000479). We thank our colleagues S.V. Babenkov, H.-C. Wu, S.N. Molotkov, D. Marchenko, A. Varykhalov, A.A. Zakharov, B.E. Murphy, A. Locatelli, T.O. Menteş, S.L. Molodtsov, D.V. Potorochin, J. Buck, O. Seeck, M. Hoesch, and J. Viefhaus for fruitful discussions and help in the preparation of this manuscript.

Conflict of interest

The authors declare that there is no conflict of interest.

Author details

Olga V. Molodtsova[1], Alexander N. Chaika[2]* and Victor Yu. Aristov[1,2]

1 Deutsches Elektronen-Synchrotron DESY, Hamburg, Germany

2 Institute of Solid State Physics of the Russian Academy of Sciences, Russian Federation

*Address all correspondence to: chaika@issp.ac.ru

References

[1] Novoselov KS, Geim AK, Morozov SV, et al. Electric field effect in atomically thin carbon films. Science. 2004;**306**:666-669. DOI: 10.1126/science.1102896

[2] Geim AK, Novoselov KS. The rise of graphene. Nature Materials. 2007;**6**: 183-191. DOI: 10.1038/nmat1849

[3] Zhang Y, Tan Y-W, Stormer HL, Kim P. Experimental observation of the quantum Hall effect and Berry's phase in graphene. Nature. 2005;**438**:201-204. DOI: 10.1038/nature04235

[4] Novoselov KS, Geim AK, Morozov SV, et al. Two-dimensional gas of massless Dirac fermions in graphene. Nature. 2005;**438**:197-200. DOI: 10.1038/nature04233

[5] Han W, Kawakami RK, Gmitra M, Fabian J. Graphene spintronics. Nature Nanotechnology. 2014;**9**:794-807. DOI: 10.1038/nnano.2014.214

[6] Yang T-Y, Balakrishnan J, Volmer F, et al. Observation of long spin-relaxation times in bilayer graphene at room temperature. Physical Review Letters. 2011;**107**:047206. DOI: 10.1103/PhysRevLett.107.047206

[7] Dlubak B, Martin M-B, Deranlot C, et al. Highly efficient spin transport in epitaxial graphene on SiC. Nature Physics. 2012;**8**:557-561. DOI: 10.1038/nphys2331

[8] Novoselov KS. Nobel lecture: Graphene: Materials in the Flatland. Reviews of Modern Physics. 2011;**83**:837-849. DOI: 10.1103/RevModPhys.83.837

[9] Novoselov KS, Fal′ko VI, Colombo L, Gellert PR, Schwab MG, Kim K. A roadmap for graphene. Nature. 2012;**490**:192-200. DOI: 10.1038/nature11458

[10] Chen Z, Lin Y-M, Rooks MJ, Avouris P. Graphene nano-ribbon electronics. Physica E (Low-dimensional Systems and Nanostructures). 2007;**40**:228-232. DOI: 10.1016/j.physe.2007.06.020

[11] de Heer WA, Berger C, Wu X, et al. Epitaxial graphene. Solid State Communications. 2007;**143**:92-100. DOI: 10.1016/j.ssc.2007.04.023

[12] Geim AK. Graphene: Status and prospects. Science. 2009;**324**:1530-1534. DOI: 10.1126/science.1158877

[13] Berger C. Electronic confinement and coherence in patterned epitaxial graphene. Science. 2006;**312**:1191-1196. DOI: 10.1126/science.1125925

[14] Hernandez Y, Nicolosi V, Lotya M, et al. High-yield production of graphene by liquid-phase exfoliation of graphite. Nature Nanotechnology. 2008;**3**: 563-568. DOI: 10.1038/nnano.2008.215

[15] Bolotin KI, Sikes KJ, Jiang Z, et al. Ultrahigh electron mobility in suspended graphene. Solid State Communications. 2008;**146**:351-355. DOI: 10.1016/j.ssc.2008.02.024

[16] Du X, Skachko I, Barker A, Andrei EY. Approaching ballistic transport in suspended graphene. Nature Nanotechnology. 2008;**3**:491-495. DOI: 10.1038/nnano.2008.199

[17] Jiao L, Zhang L, Wang X, Diankov G, Dai H. Narrow graphene nanoribbons from carbon nanotubes. Nature. 2009;**458**:877-880. DOI: 10.1038/nature07919

[18] Choucair M, Thordarson P, Stride JA. Gram-scale production of graphene based on solvothermal synthesis and sonication. Nature Nanotechnology. 2009;**4**:30-33. DOI: 10.1038/nnano.2008.365

[19] Chakrabarti A, Lu J, Skrabutenas JC, et al. Conversion of carbon dioxide to few-layer graphene. Journal of Materials Chemistry. 2011;**21**:9491. DOI: 10.1039/c1jm11227a

[20] Rümmeli MH, Gorantla S, Bachmatiuk A, et al. On the role of vapor trapping for chemical vapor deposition (CVD) grown graphene over copper. Chemistry of Materials. 2013;**25**:4861-4866. DOI: 10.1021/cm401669k

[21] Li X, Cai W, An J, et al. Large-area synthesis of high-quality and uniform graphene films on copper foils. Science. 2009;**324**:1312-1314. DOI: 10.1126/science.1171245

[22] Rümmeli MH, Bachmatiuk A, Scott A, et al. Direct low-temperature nanographene CVD synthesis over a dielectric insulator. ACS Nano. 2010;**4**:4206-4210. DOI: 10.1021/nn100971s

[23] Wei D, Xu X. Laser direct growth of graphene on silicon substrate. Applied Physics Letters. 2012;**100**:023110. DOI: 10.1063/1.3675636

[24] Michon A, Tiberj A, Vézian S, et al. Graphene growth on AlN templates on silicon using propane-hydrogen chemical vapor deposition. Applied Physics Letters. 2014;**104**:071912. DOI: 10.1063/1.4866285

[25] Chen J, Wen Y, Guo Y, et al. Oxygen-aided synthesis of polycrystalline graphene on silicon dioxide substrates. Journal of the American Chemical Society. 2011;**133**:17548-17551. DOI: 10.1021/ja2063633

[26] Wang G, Zhang M, Zhu Y, et al. Direct growth of graphene film on germanium substrate. Scientific Reports. 2013;**3**:2465. DOI: 10.1038/srep02465

[27] Tang S, Ding G, Xie X, et al. Nucleation and growth of single crystal graphene on hexagonal boron nitride. Carbon N Y. 2012;**50**:329-331. DOI: 10.1016/j.carbon.2011.07.062

[28] Hwang J, Kim M, Campbell D, et al. van der Waals epitaxial growth of graphene on sapphire by chemical vapor deposition without a metal catalyst. ACS Nano. 2013;**7**:385-395. DOI: 10.1021/nn305486x

[29] Ismach A, Druzgalski C, Penwell S, et al. Direct chemical vapor deposition of graphene on dielectric surfaces. Nano Letters. 2010;**10**:1542-1548. DOI: 10.1021/nl9037714

[30] Pasternak I, Wesolowski M, Jozwik I, et al. Graphene growth on Ge(100)/Si(100) substrates by CVD method. Scientific Reports. 2016;**6**:21773. DOI: 10.1038/srep21773

[31] Van Bommel AJ, Crombeen JE, Van Tooren A. LEED and Auger electron observations of the SiC(0001) surface. Surface Science. 1975;**48**:463-472. DOI: 10.1016/0039-6028(75)90419-7

[32] Forbeaux I, Themlin J-M, Debever J-M. Heteroepitaxial graphite on 6H-SiC(0001): Interface formation through conduction-band electronic structure. Physical Review B. 1998;**58**:16396-16406. DOI: 10.1103/PhysRevB.58.16396

[33] Virojanadara C, Syväjarvi M, Yakimova R, Johansson LI, Zakharov AA, Balasubramanian T. Homogeneous large-area graphene layer growth on 6H-SiC(0001). Physical Review B. 2008;**78**:245403. DOI: 10.1103/PhysRevB.78.245403

[34] Emtsev KV, Bostwick A, Horn K, et al. Towards wafer-size graphene layers by atmospheric pressure graphitization of silicon carbide. Nature Materials. 2009;**8**:203-207. DOI: 10.1038/nmat2382

[35] Ohta T, Bostwick A, Seyller T, Horn K, Rotenberg E. Controlling the

electronic structure of bilayer graphene. Science. 2006;**313**:951-954. DOI: 10.1126/science.1130681

[36] Riedl C, Starke U, Bernhardt J, Franke M, Heinz K. Structural properties of the graphene-SiC(0001) interface as a key for the preparation of homogeneous large-terrace graphene surfaces. Physical Review B. 2007;**76**:245406. DOI: 10.1103/PhysRevB.76.245406

[37] First PN, de Heer WA, Seyller T, Berger C, Stroscio JA, Moon J-S. Epitaxial graphenes on silicon carbide. MRS Bulletin. 2010;**35**:296-305. DOI: 10.1557/mrs2010.552

[38] Berger C, Song Z, Li T, et al. Ultrathin epitaxial graphite: 2D electron gas properties and a route toward graphene-based nanoelectronics. The Journal of Physical Chemistry. B. 2004;**108**:19912-19916. DOI: 10.1021/jp040650f

[39] Hass J, de Heer WA, Conrad EH. The growth and morphology of epitaxial multilayer graphene. Journal of Physics. Condensed Matter. 2008;**20**:323202. DOI: 10.1088/0953-8984/20/32/323202

[40] de Heer WA, Berger C, Wu X, et al. Epitaxial graphene electronic structure and transport. Journal of Physics D: Applied Physics. 2010;**43**:374007. DOI: 10.1088/0022-3727/43/37/374007

[41] Tejeda A, Taleb-Ibrahimi A, de Heer W, Berger C, Conrad EH. Electronic structure of epitaxial graphene grown on the C-face of SiC and its relation to the structure. New Journal of Physics. 2012;**14**:125007. DOI: 10.1088/1367-2630/14/12/125007

[42] Sprinkle M, Siegel D, Hu Y, et al. First direct observation of a nearly ideal graphene band structure. Physical Review Letters. 2009;**103**:226803. DOI: 10.1103/PhysRevLett.103.226803

[43] Nishino S, Powell JA, Will HA. Production of large-area single-crystal wafers of cubic SiC for semiconductor devices. Applied Physics Letters. 1983;**42**:460-462. DOI: 10.1063/1.93970

[44] Feng ZC, Mascarenhas AJ, Choyke WJ, Powell JA. Raman scattering studies of chemical-vapor-deposited cubic SiC films of (100) Si. Journal of Applied Physics. 1988;**64**:3176-3186. DOI: 10.1063/1.341533

[45] Shigeta M, Fujii Y, Furukawa K, Suzuki A, Nakajima S. Chemical vapor deposition of single-crystal films of cubic SiC on patterned Si substrates. Applied Physics Letters. 1989;**55**:1522-1524. DOI: 10.1063/1.102252

[46] Golecki I, Reidinger F, Marti J. Single-crystalline, epitaxial cubic SiC films grown on (100) Si at 750°C by chemical vapor deposition. Applied Physics Letters. 1992;**60**:1703-1705. DOI: 10.1063/1.107191

[47] Coletti C, Frewin CL, Saddow SE, Hetzel M, Virojanadara C, Starke U. Surface studies of hydrogen etched 3C-SiC(001) on Si(001). Applied Physics Letters. 2007;**91**:061914. DOI: 10.1063/1.2768870

[48] Miyamoto Y, Handa H, Saito E, et al. Raman-scattering spectroscopy of epitaxial graphene formed on SiC film on Si substrate. e-Journal of Surface Science and Nanotechnology. 2009;7:107-109. DOI: 10.1380/ejssnt.2009.107

[49] Suemitsu M, Miyamoto Y, Handa H, Konno A. Graphene formation on a 3C-SiC(111) thin film grown on Si(110) substrate. e-Journal of Surface Science and Nanotechnology. 2009;7:311-313. DOI: 10.1380/ejssnt.2009.311

[50] Aristov VY, Urbanik G, Kummer K, et al. Graphene synthesis on cubic SiC/Si wafers. Perspectives for mass production of graphene-based

electronic devices. Nano Letters. 2010;**10**:992-995. DOI: 10.1021/nl904115h

[51] Suemitsu M, Fukidome H. Epitaxial graphene on silicon substrates. Journal of Physics D: Applied Physics. 2010;**43**:374012. DOI: 10.1088/0022-3727/43/37/374012

[52] Fukidome H, Miyamoto Y, Handa H, Saito E, Suemitsu M. Epitaxial growth processes of graphene on silicon substrates. Japanese Journal of Applied Physics. 2010;**49**:01AH03. DOI: 10.1143/JJAP.49.01AH03

[53] Ouerghi A, Kahouli A, Lucot D, et al. Epitaxial graphene on cubic SiC(111)/Si(111) substrate. Applied Physics Letters. 2010;**96**:191910. DOI: 10.1063/1.3427406

[54] Ouerghi A, Belkhou R, Marangolo M, et al. Structural coherency of epitaxial graphene on 3C–SiC(111) epilayers on Si(111). Applied Physics Letters. 2010;**97**:161905. DOI: 10.1063/1.3497287

[55] Abe S, Handa H, Takahashi R, Imaizumi K, Fukidome H, Suemitsu M. Surface chemistry involved in epitaxy of graphene on 3C-SiC(111)/Si(111). Nanoscale Research Letters. 2010;**5**:1888-1891. DOI: 10.1007/s11671-010-9731-x

[56] Ouerghi A, Marangolo M, Belkhou R, et al. Epitaxial graphene on 3C-SiC(111) pseudosubstrate: Structural and electronic properties. Physical Review B. 2010;**82**:125445. DOI: 10.1103/PhysRevB.82.125445

[57] Coletti C, Emtsev KV, Zakharov AA, Ouisse T, Chaussende D, Starke U. Large area quasi-free standing monolayer graphene on 3C-SiC(111). Applied Physics Letters. 2011;**99**:081904. DOI: 10.1063/1.3618674

[58] Takahashi R, Handa H, Abe S, et al. Low-energy-electron-diffraction and X-ray-phototelectron-spectroscopy studies of graphitization of 3C-SiC(111) thin film on Si(111) substrate. Japanese Journal of Applied Physics. 2011;**50**:070103. DOI: 10.1143/JJAP.50.070103

[59] Handa H, Takahashi R, Abe S, et al. Transmission electron microscopy and Raman-scattering spectroscopy observation on the interface structure of graphene formed on Si substrates with various orientations. Japanese Journal of Applied Physics. 2011;**50**:04DH02. DOI: 10.1143/JJAP.50.04DH02

[60] Otsuji T, Boubanga Tombet SA, Satou A, et al. Graphene-based devices in terahertz science and technology. Journal of Physics D: Applied Physics. 2012;**45**:303001. DOI: 10.1088/0022-3727/45/30/303001

[61] Portail M, Michon A, Vézian S, et al. Growth mode and electric properties of graphene and graphitic phase grown by argon–propane assisted CVD on 3C–SiC/Si and 6H–SiC. Journal of Crystal Growth. 2012;**349**:27-35. DOI: 10.1016/j.jcrysgro.2012.04.004

[62] Starke U, Coletti C, Emtsev K, Zakharov AA, Ouisse T, Chaussende D. Large area quasi-free standing monolayer graphene on 3C-SiC(111). Materials Science Forum. 2012;**717-720**:617-620. DOI: 10.4028/www.scientific.net/MSF.717-720.617

[63] Coletti C, Forti S, Principi A, et al. Revealing the electronic band structure of trilayer graphene on SiC: An angle-resolved photoemission study. Physical Review B. 2013;**88**:155439. DOI: 10.1103/PhysRevB.88.155439

[64] Darakchieva V, Boosalis A, Zakharov AA, et al. Large-area microfocal spectroscopic ellipsometry mapping of thickness and electronic properties of epitaxial graphene on Si- and C-face of 3C-SiC(111). Applied

Physics Letters. 2013;**102**:213116. DOI: 10.1063/1.4808379

[65] Aryal HR, Fujita K, Banno K, Egawa T. Epitaxial graphene on Si(111) substrate grown by annealing 3C-SiC/carbonized silicon. Japanese Journal of Applied Physics. 2012;**51**:01AH05. DOI: 10.1143/JJAP.51.01AH05

[66] Fukidome H, Abe S, Takahashi R, et al. Controls over structural and electronic properties of epitaxial graphene on silicon using surface termination of 3C-SiC(111)/Si. Applied Physics Express. 2011;**4**:115104. DOI: 10.1143/APEX.4.115104

[67] Sanbonsuge S, Abe S, Handa H, et al. Improvement in film quality of epitaxial graphene on SiC(111)/Si(111) by SiH_4 pretreatment. Japanese Journal of Applied Physics. 2012;**51**:06FD10. DOI: 10.1143/JJAP.51.06FD10

[68] Hsia B, Ferralis N, Senesky DG, Pisano AP, Carraro C, Maboudian R. Epitaxial graphene growth on 3C–SiC(111)/AlN(0001)/Si(100). Electrochemical and Solid-State Letters. 2011;**14**:K13. DOI: 10.1149/1.3518713

[69] Jiao S, Murakami Y, Nagasawa H, et al. High quality graphene formation on 3C-SiC/4H-AlN/Si heterostructure. Materials Science Forum. 2014;**806**: 89-93. DOI: 10.4028/www.scientific.net/MSF.806.89

[70] Gupta B, Notarianni M, Mishra N, Shafiei M, Iacopi F, Motta N. Evolution of epitaxial graphene layers on 3C SiC/Si (111) as a function of annealing temperature in UHV. Carbon N Y. 2014;**68**:563-572. DOI: 10.1016/j.carbon.2013.11.035

[71] Gupta B, Placidi E, Hogan C, Mishra N, Iacopi F, Motta N. The transition from 3C SiC(111) to graphene captured by Ultra High Vacuum Scanning Tunneling Microscopy. Carbon N Y. 2015;**91**:378-385. DOI: 10.1016/j.carbon.2015.05.011

[72] Pierucci D, Sediri H, Hajlaoui M, et al. Evidence for flat bands near the fermi level in epitaxial rhombohedral multilayer graphene. ACS Nano. 2015;**9**:5432-5439. DOI: 10.1021/acsnano.5b01239

[73] Zarotti F, Gupta B, Iacopi F, Sgarlata A, Tomellini M, Motta N. Time evolution of graphene growth on SiC as a function of annealing temperature. Carbon N Y. 2016;**98**:307-312. DOI: 10.1016/j.carbon.2015.11.026

[74] Gupta B, Di Bernardo I, Mondelli P, et al. Effect of substrate polishing on the growth of graphene on 3C-SiC(111)/Si(111) by high temperature annealing. Nanotechnology. 2016;**27**:185601. DOI: 10.1088/0957-4484/27/18/185601

[75] Shi Y, Zakharov AA, Ivanov IG, et al. Elimination of step bunching in the growth of large-area monolayer and multilayer graphene on off-axis 3C-SiC (111). Carbon N Y. 2016;**140**:533-542. DOI: 10.1016/j.carbon.2018.08.042

[76] Bouhafs C, Stanishev V, Zakharov AA, et al. Decoupling and ordering of multilayer graphene on C-face 3C-SiC(111). Applied Physics Letters. 2016;**109**:203102. DOI: 10.1063/1.4967525

[77] Sambonsuge S, Jiao S, Nagasawa H, et al. Formation of qualified epitaxial graphene on Si substrates using two-step heteroexpitaxy of C-terminated 3C-SiC(-1-1-1) on Si(110). Diamond and Related Materials. 2016;**67**:51-53. DOI: 10.1016/j.diamond.2016.02.020

[78] Nemec L, Lazarevic F, Rinke P, et al. Why graphene growth is very different on the C face than on the Si face of SiC: Insights from surface equilibria and the (3×3)-3C-SiC(-1-1-1) reconstruction. Physical Review B. 2015;**91**:161408. DOI: 10.1103/PhysRevB.91.161408

[79] Mondelli P, Gupta B, Betti MG, et al. High quality epitaxial graphene by hydrogen-etching of 3C-SiC(111) thin-film on Si(111). Nanotechnology. 2017;**28**:115601. DOI: 10.1088/1361-6528/aa5a48

[80] Amjadipour M, MacLeod J, Lipton-Duffin J, et al. Epitaxial graphene growth on FIB patterned 3C-SiC nanostructures on Si (111): Reducing milling damage. Nanotechnology. 2017;**28**:345602. DOI: 10.1088/1361-6528/aa752e

[81] Amjadipour M, Tadich A, Boeckl JJ, et al. Quasi free-standing epitaxial graphene fabrication on 3C-SiC/Si(111). Nanotechnology. 2018;**29**:145601. DOI: 10.1088/1361-6528/aaab1a

[82] Ouerghi A, Ridene M, Balan A, et al. Sharp interface in epitaxial graphene layers on 3C-SiC(100)/Si(100) wafers. Physical Review B. 2011;**83**:205429. DOI: 10.1103/PhysRevB.83.205429

[83] Gogneau N, Balan A, Ridene M, Shukla A, Ouerghi A. Control of the degree of surface graphitization on 3C-SiC(100)/Si(100). Surface Science. 2012;**606**:217-220. DOI: 10.1016/j.susc.2011.09.021

[84] Ouerghi A, Balan A, Castelli C, et al. Epitaxial graphene on single domain 3C-SiC(100) thin films grown on off-axis Si(100). Applied Physics Letters. 2012;**101**:021603. DOI: 10.1063/1.4734396

[85] Chaika AN, Molodtsova OV, Zakharov AA, et al. Continuous wafer-scale graphene on cubic-SiC(001). Nano Research. 2013;**6**:562-570. DOI: 10.1007/s12274-013-0331-9

[86] Abe S, Handa H, Takahashi R, Imaizumi K, Fukidome H, Suemitsu M. Temperature-programmed desorption observation of graphene-on-silicon process. Japanese Journal of Applied Physics. 2011;**50**:070102. DOI: 10.1143/JJAP.50.070102

[87] Chaika AN, Molodtsova OV, Zakharov AA, et al. Rotated domain network in graphene on cubic-SiC(001). Nanotechnology. 2014;**25**:135605. DOI: 10.1088/0957-4484/25/13/135605

[88] Velez-Fort E, Silly MG, Belkhou R, Shukla A, Sirotti F, Ouerghi A. Edge state in epitaxial nanographene on 3C-SiC(100)/Si(100) substrate. Applied Physics Letters. 2013;**103**:083101. DOI: 10.1063/1.4818547

[89] Gogneau N, Ben Gouider Trabelsi A, et al. Investigation of structural and electronic properties of epitaxial graphene on 3C-SiC(100)/Si(100) substrates. Nanotechnology, Science and Applications. 2014;**7**:85. DOI: 10.2147/NSA.S60324

[90] Hens P, Zakharov AA, Iakimov T, Syväjärvi M, Yakimova R. Large area buffer-free graphene on non-polar (001) cubic silicon carbide. Carbon N Y. 2014;**80**:823-829. DOI: 10.1016/j.carbon.2014.09.041

[91] Suemitsu M, Jiao S, Fukidome H, Tateno Y, Makabe I, Nakabayashi T. Epitaxial graphene formation on 3C-SiC/Si thin films. Journal of Physics D: Applied Physics. 2014;**47**:094016. DOI: 10.1088/0022-3727/47/9/094016

[92] Iacopi F, Mishra N, Cunning BV, et al. A catalytic alloy approach for graphene on epitaxial SiC on silicon wafers. Journal of Materials Research. 2015;**30**:609-616. DOI: 10.1557/jmr.2015.3

[93] Wu H-C, Chaika AN, Huang T-W, et al. Transport gap opening and high on-off current ratio in trilayer graphene with self-aligned nanodomain boundaries. ACS Nano. 2015;**9**:8967-8975. DOI: 10.1021/acsnano.5b02877

[94] Huang H, Liang Wong S, Tin C-C, et al. Epitaxial growth and characterization of graphene on free-standing polycrystalline 3C-SiC. Journal

of Applied Physics. 2011;**110**:014308. DOI: 10.1063/1.3602993

[95] Ide T, Kawai Y, Handa H, et al. Epitaxy of graphene on 3C-SiC(111) thin films on microfabricated Si(111) substrates. Japanese Journal of Applied Physics. 2012;**51**:06FD02. DOI: 10.1143/JJAP.51.06FD02

[96] Cunning BV, Ahmed M, Mishra N, Kermany AR, Wood B, Iacopi F. Graphitized silicon carbide microbeams: Wafer-level, self-aligned graphene on silicon wafers. Nanotechnology. 2014;**25**:325301. DOI: 10.1088/0957-4484/25/32/325301

[97] Fukidome H, Kawai Y, Fromm F, et al. Precise control of epitaxy of graphene by microfabricating SiC substrate. Applied Physics Letters. 2012;**101**:041605. DOI: 10.1063/1.4740271

[98] Bantaculo R, Fukidome H, Suemitsu M. Correlation between the residual stress in 3C-SiC/Si epifilm and the quality of epitaxial graphene formed thereon. IOP Conference Series Materials Science and Engineering. 2015;**79**:012004. DOI: 10.1088/1757-899X/79/1/012004

[99] Yazdi GR, Vasiliauskas R, Iakimov T, Zakharov A, Syväjärvi M, Yakimova R. Growth of large area monolayer graphene on 3C-SiC and a comparison with other SiC polytypes. Carbon N Y. 2013;**57**:477-484. DOI: 10.1016/j.carbon.2013.02.022

[100] Fukidome H, Ide T, Kawai Y, et al. Microscopically-tuned band structure of epitaxial graphene through interface and stacking variations using Si substrate microfabrication. Scientific Reports. 2015;**4**:5173. DOI: 10.1038/srep05173

[101] Wu H-C, Chaika AN, Hsu M-C, et al. Large positive in-plane magnetoresistance induced by localized states at nanodomain boundaries in graphene. Nature Communications. 2017;**8**:14453. DOI: 10.1038/ncomms14453

[102] Aristov VY, Chaika AN, Molodtsova OV, et al. Layer-by-layer graphene growth on β-SiC/Si(001). ACS Nano. 2019;**13**:526-535. DOI: 10.1021/acsnano.8b07237

[103] Chaika AN, Aristov VY, Molodtsova OV. Graphene on cubic-SiC. Progress in Materials Science. 2017;**89**:1-30. DOI: 10.1016/j.pmatsci.2017.04.010

[104] Fasolino A, Los JH, Katsnelson MI. Intrinsic ripples in graphene. Nature Materials. 2007;**6**:858-861. DOI: 10.1038/nmat2011

[105] Meyer JC, Geim AK, Katsnelson MI, Novoselov KS, Booth TJ, Roth S. The structure of suspended graphene sheets. Nature. 2007;**446**:60-63. DOI: 10.1038/nature05545

[106] Hibino H, Kageshima H, Maeda F, Nagase M, Kobayashi Y, Yamaguchi H. Microscopic thickness determination of thin graphite films formed on SiC from quantized oscillation in reflectivity of low-energy electrons. Physical Review B. 2008;**77**:075413. DOI: 10.1103/PhysRevB.77.075413

[107] Riedl C, Coletti C, Iwasaki T, Zakharov AA, Starke U. Quasi-free-standing epitaxial graphene on SiC obtained by hydrogen intercalation. Physical Review Letters. 2009;**103**:246804. DOI: 10.1103/PhysRevLett.103.246804

[108] Shirley EL, Terminello LJ, Santoni A, Himpsel FJ. Brillouin-zone-selection effects in graphite photoelectron angular distributions. Physical Review B. 1995;**51**:13614-13622. DOI: 10.1103/PhysRevB.51.13614

[109] Babenkov SV, Aristov VY, Molodtsova OV, et al. A new

dynamic-XPS end-station for beamline PO_4 at PETRA III/DESY. Nuclear Instruments and Methods in Physics Research Section A: Accelerators, Spectrometers, Detectors and Associated Equipment. 2015;**777**:189-193. DOI: 10.1016/j.nima.2014.12.065

[110] Aristov VY. β-SiC(100) surface: Atomic structures and electronic properties. Physics-Uspekhi. 2001;**44**:761-783. DOI: 10.1070/PU2001v044n08ABEH000979

[111] Soukiassian PG, Enriquez HB. Atomic scale control and understanding of cubic silicon carbide surface reconstructions, nanostructures and nanochemistry. Journal of Physics. Condensed Matter. 2004;**16**:S1611-S1658. DOI: 10.1088/0953-8984/16/17/011

[112] Hupalo M, Conrad EH, Tringides MC. Growth mechanism for epitaxial graphene on vicinal 6 H-SiC(0001) surfaces: A scanning tunneling microscopy study. Physical Review B. 2009;**80**:041401. DOI: 10.1103/PhysRevB.80.041401

[113] Wang Q, Zhang W, Wang L, He K, Ma X, Xue Q. Large-scale uniform bilayer graphene prepared by vacuum graphitization of 6H-SiC(0001) substrates. Journal of Physics. Condensed Matter. 2013;**25**:095002. DOI: 10.1088/0953-8984/25/9/095002

[114] Patera LL, Bianchini F, Africh C, et al. Real-time imaging of adatom-promoted graphene growth on nickel. Science. 2018;**359**:1243-1246. DOI: 10.1126/science.aan8782

[115] Yazyev OV, Louie SG. Electronic transport in polycrystalline graphene. Nature Materials. 2010;**9**:806-809. DOI: 10.1038/nmat2830

[116] Huertas-Hernando D, Guinea F, Brataas A. Spin-orbit coupling in curved graphene, fullerenes, nanotubes, and nanotube caps. Physical Review B. 2006;**74**:155426. DOI: 10.1103/PhysRevB.74.155426

Chapter 8

Ceramics (Si- and Al-Based Oxides)-Graphene Hybrids and Advanced Applications

Mujtaba Ikram and Muhammad Umer Farooq

Abstract

This book chapter will describe the recent advancements in advanced carbon-ceramics based hybrid materials, enhanced properties and efficient applications. There are various fabrication methods, Firstly, authors will discuss a solvothermal/hot-pressing method which is employed to fabricate hybrids composed of cross-linked γ-Al_2O_3 nanorods and reduced graphite oxide (rGO) platelets. After calcination and hot-press processing, monoliths of Al_2O_3-rGO hybrids are obtained with improved physical properties. It is found that the oxygen-containing groups on graphene oxide benefit to the adsorption of Al-$(OC_3H_7)_3$ (aluminum isopropoxide), leading to the uniform dispersion of rGO with Al_2O_3 which is hydrolyzed from aluminum isopropoxide in solvothermal reactions. Further, this research methodology has been extended to another ceramics-graphene nanostructure assembly, i.e., silica-rGO hybrids, by optimizing experimental conditions for the hydrothermal and hot pressing process. This book chapter will be a significant contribution for the applications of ceramics-graphene assembly nanomaterials, which can be made by simple fabrication route and which can be further applied as electrolytes, catalysts, conductive, electrochemically active, and as dielectric materials for the high-temperature applications due to enhanced physical properties.

Keywords: graphene, ceramics, enhanced physical properties, solvothermal, hot pressing, calcination, hydrolyzed, platelets, reduced graphene oxide, thermal properties, electrical properties, dielectric properties, mechanical properties

1. Introduction

In recent years, hybrid nanostructures have achieved great technological interest, because of the ability of the nanostructures which intends to design and manufacture hybrid materials with enhanced physical properties [1, 2]. For a desired technological application, various excellent properties can be combined into a single hybrid nanostructure, this is an excellent feature of the hybrid nanostructures. Moreover, in hybrid nanostructures, the enhanced electronic interactions among various types of components are possible, which is the main reason to enhance the physical properties of hybrids [3, 4]. In ceramics-graphene hybrids, enhanced physical properties can be used in the high electrical, mechanical, and thermal applications, because of higher electrical conductivity, thermal conductivity, dielectric and better mechanical properties [5]. Due to the versatile physical and chemical properties [6, 7], graphene has opted in huge scientific activities in progress of research and development for the field of science and technology. The combination of suitable electrical, mechanical, thermal, and physical properties can apply not only for a broad spectrum of applications but also as a basic essential unite for the fundamental and advanced technological research [8, 9]. In hybrids or composite, scientists worldwide have been impressed by the unique combination of individual physical properties of graphene [8], that makes graphene an ideal option, or as an advanced component in both the hybrids as well as composites. For monolithic ceramics, in particular, physical properties of graphene have been researched and investigated as an ideal component in the monolithic ceramics-graphene hybrids or composite [10, 11]. Due to the higher strength, stiffness, and high-temperature stability, monolithic ceramics are used and known commonly as a very promising structural material for mechanical and high-temperature applications [12, 13]. But the field of application is still vacant due to mechanical unreliability, and very lower electrical conductivity, and limited physical properties of the monolithic ceramics.

Due to graphene's extraordinary physical properties [14, 15], incorporating graphene in the ceramics can have great potential for electro-conductive, and high mechanical applications. In graphene-ceramics hybrids, enhanced physical properties could be implemented in a wide range of the material related applications in the field of aerospace, processing industries, and military based applications. [16, 17] In view of the fabrication routes [17, 18], development of the graphene ceramic hybrids is still complicated due to reinforcement particle at a very nanometric scale. For fabrication methodology, practical issues can be classified into wide categories such as (1) in ceramic nanostructures, homogenous dispersion of the graphene is important for enhanced physical properties; (2) easy processing route of graphene-ceramics hybrids or nanocomposites is necessary; (3) interfacial bonding and interaction between graphene and ceramic nanostructures are very important as it directly reduces the physical properties of the graphene-ceramics hybrids or nanocomposites. For ceramics-graphene hybrids, uniform dispersion of the graphene in the ceramic matrices is an important factor [12]. Due to the high surface area of graphene [8], proper graphene dispersion is a very important factor, which further ensures efficient load transfer between graphene and available ceramics nanostructures in the hybrids. This is major concerns during the incorporation of graphene in the ceramics, due to the higher surface area of graphene [6]. For this purpose, scientists have used various dispersing agents [19, 20]. The use of dispersing agent gives rise to higher surface potential, double layer formation, as well possibility of strong electrostatic repulsion, which helps to uniform dispersion of graphene in graphene-ceramics hybrids. Mechanical dispersion of graphene is possible through many routes such as ultra-sonication, ball milling, and stirring [7, 21].

The enhanced physical properties of graphene-ceramic materials depend upon many factors such as thin layers of graphene, fine particles size and phase homogeneity [22]. For graphene-ceramics hybrids, well aligned and controlled nanostructures are important in toughening of hybrids. In literature, there are available many fabrication ways of graphene-ceramics hybrids such as powder processing, colloidal processing and sol-gel fabrication. In most of the work on graphene-ceramics by conventional powder routes, physical properties are not as good as expected because graphene is prone to agglomeration due to van der Waals forces. Therefore, in this chapter, our focus is on new solvothermal-hot press method, which is used to fabricate alumina-rGO, and silica-rGO hybrids, with a systematic study on enhanced physical properties of the hybrids for efficient application. In hybrids, the physical properties are enhanced by a great degree, because of use of calcination conditions, as well as the hot-pressing conditions. The two structural ceramics, which we will discuss in this chapter, are alumina and silica, respectively.

Ceramics usually have a brittle attribute with low strength [11]. Among many ceramics, alumina is one of the widely used structural ceramic due to the shaping capability, and the good thermal conductivity [12]. Alumina has applications in the field of high-speed cutting tools, dental implants and insulators [13, 14]. To improve the mechanical properties, carbon nanotubes have been used to enhance the fracture toughness (by a degree of 94%), hardness (by a degree of 13%), and flexural strength (by a degree of 6.4%) of the alumina [15]. Ball-milled alumina/zirconia/graphene composite has been investigated with 40% enhanced fracture toughness by adding the graphene platelets [16]. In another work, alumina-rGO nanocomposites have been fabricated by the dry sol-gel method, from which it was indicated that BET surface area of rGO is essential to enhance the surface charge properties of hybrids [17]. In another work, alumina-graphene composite films have been reported with low optical gap (1.53 eV) [18]. Alumina-rGO nanocomposite by in-situ deposition have shown morphologies of nanoparticles of alumina structures on rGO with BET surface area of 242.4 $m^2\ g^{-1}$ and the low porosity [19]. Alumina/rGO/poly(ethylenimine) composite has been used to capture carbon dioxide from the flue gas [20]. In a microwave preparation of alumina-rGO composites, the grain size of alumina matrix was reduced to 180 nm compared to 475 nm of the conventional sintering process, leading to an increase in Young's modulus of 180 from 148 GPa under the same measurement condition [21]. In this chapter, we will discuss the preparation of hybrids consisting of γ-Al_2O_3 nanorods and rGO by a solvothermal method. This solvothermal method is used to form hybrids composed of cross-linked γ-Al_2O_3 nanorods and reduced graphite oxide (rGO) platelets. With further hot pressing, a hybrid monolith has been made for the systematic study on enhanced physical properties of the hybrids.

The second structural ceramics, which we will discuss in this chapter, is silica. Among various kinds of the ceramics, silica particles are one of the widely used additive ceramic due to its functionalized ability and stability with a range of materials [23, 24]. Silica has various applications in the domain of polymer, biomedical, and composite engineering [25]. To improve the physical properties, rGO has been used to enhance thermal conductivity (0.452 $Wm^{-1}\ K^{-1}$), storage modulus (3.56) and dielectric constant (77.23) of epoxy/SiO_2/rGO hybrid [26]. In another work, in-situ sol-gel processed silica nanoparticles decorated with graphene oxide sheets are obtained, from which it was found that presence of rGO is essential to improve the corrosion resistance, dispersion, and the barrier properties of hybrid [27]. In another work, SiO_2-graphene hybrids have shown superior gas sensing response (31.5%) towards 50 ppm NH3 for 850 s, in comparison to rGO based sensor (1.5%) [28]. Using one-step hydrothermal method, SiO_2-rGO nanohybrid has shown comparatively better BET surface area (676 $m^2\ g^{-1}$) and 98.8% Cr(VI) adsorption efficiency [29]. SiO_2 supported polyvinylidene fluoride has shown high dielectric constant (72.94) and low dielectric loss (0.059), which is due to the addition of ultrathin graphene [30]. Among the field of diverse inorganic particles, still, there is need to do much of scientific research for enhanced physical properties of silica-carbon based hybrids.

2. Experimental setup/fabrication route for ceramics-graphene hybrids

There are many fabrication methods for graphene-ceramics materials. Here in this chapter, In brief, the preparation of ceramics-graphene hybrids was done by mixing GO with cyclohexane and corresponding metal alkoxide followed by a solvothermal reaction. For the preparation, 0.1 g of GO was firstly dispersed in 35 ml cyclohexane, after which desired amount of corresponding metal alkoxide was added drop by drop to the GO suspension above. Centrifugation was used to separate the products which were then washed out several times with cyclohexane. The solid samples thus obtained are denoted as (Ceramics Oxide)x/GO. (Ceramic Oxide) x/GO was dispersed again in 50 ml cyclohexane and then transferred to a 100 ml Teflon-lined stainless-steel autoclave for hydrothermal reaction. After the reaction was carried out, samples denoted as (Ceramic Oxide)x/rGO. (Ceramic Oxide)x/rGO was then calcinated at a temperature above 700 K for a specific interval of time to form Ceramics Oxide/rGO hybrids. Graphene-ceramics hybrid powder containing different wt.% of rGO were obtained using the same method. Hot pressing of ceramics-graphene hybrid powder was performed in a vacuum furnace (model number OTF-1200X-VHP4). The flowchart fabrication scheme of gamma alumina-rGO hybrid with detailed experimental conditions is represented in **Figure 1**.

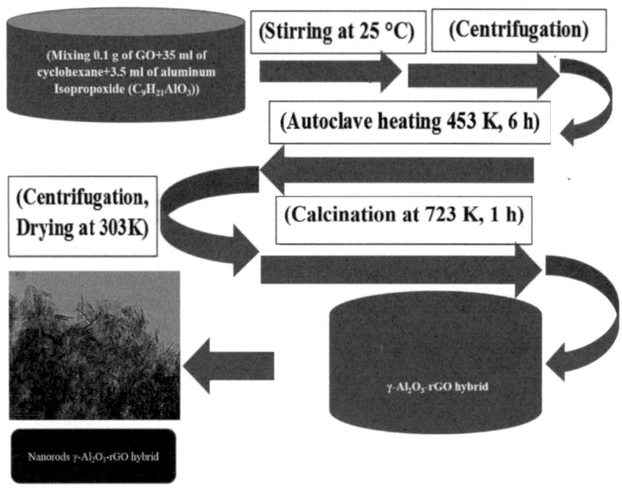

Figure 1.
Flow chart fabrication scheme for γ-Al_2O_3-rGO hybrids.

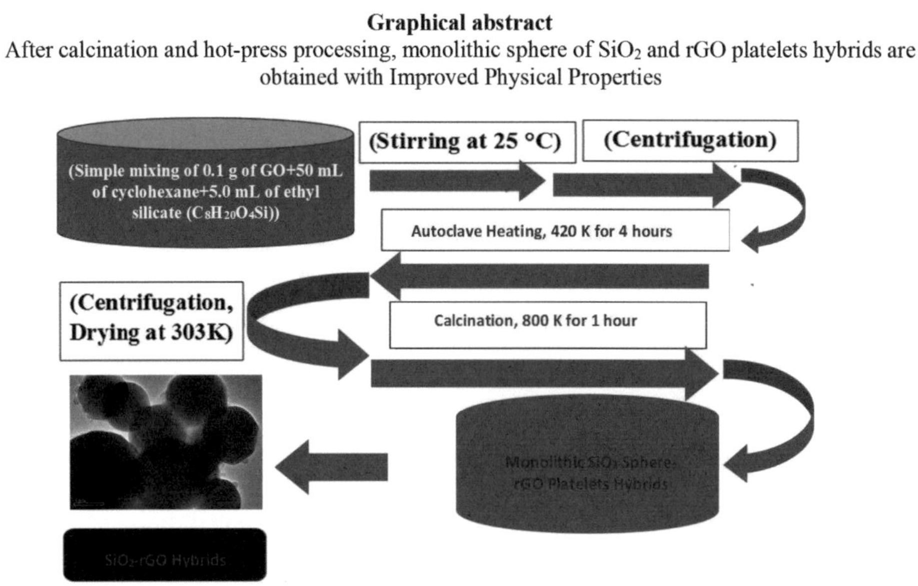

Figure 2.
Flow chart fabrication scheme for SiO_2-rGO hybrids.

Figure 2 shows the fabrication flowchart of silica-rGO hybrids, which are obtained using specific temperature and hot pressing conditions.

3. Physical properties of alumina-graphene hybrids for technological and applied applications

TGA (shown in **Figure 3**) of γ-Al_2O_3-rGO powder samples show that different calcination times has led to different concentrations of rGO in the hybrids. The TGA curves of all hybrids show a stable weight loss between 400 and 600°C, as a result of the removal of all carbon-related materials, and other impurities (if any) after heating these hybrids to 800°C in an air atmosphere. For the samples with 3-, 2- and 1-h calcination time, the 7.705, 12.830 and 16.707 wt.% loss were calculated. For 1-h sample, the unique weight loss is observed.

SEM image in (**Figure 4a**) for bare Al_2O_3 shows particles like morphology. The size of particles has ranged from 500 nm to few micrometers. TEM image in (**Figure 4b**) shows elongated nanocrystals or nanorods of bare Al_2O_3. Sample before calcination but after autoclave heating has been referred as Al(O)x/rGO. SEM image of Al(O)x/rGO after heating in an autoclave at a temperature of 453 K for 6 h but before calcination is shown in the **Figure 4c**. Even after calcination at 723 K for 2 h, the SEM image in **Figure 4d** shows the same particle like morphology but size of particles has ranged from 1 micrometers to few micrometers. TEM image of γ-Al_2O_3-rGO hybrids after calcination at 723 K for 2 h is shown in the **Figure 4e**. It shows elongated and fine nanorods of γ-Al_2O_3 with rGO layer in hybrids. The TEM image in **Figure 4e** indicates the presence of a very thin rGO layer, which acts as a continuum matrix in these hybrids.

The presence of rGO can also be confirmed by closely observing **Figure 4e**. In this figure, low-contrast features are actually edges or small portions of the graphene sheet (**Figure 4e**) on which γ-Al_2O_3 is uniformly distributed in dense concentrations. Further, the selected area electron diffraction pattern presented in **Figure 4f** shows the inter-planar spacing's of D = 0.175 nm and D = 0.151 nm, corresponding to the (200) and (111) planes of γ-Al_2O_3. The fabrication of the γ-Al_2O_3 phase was

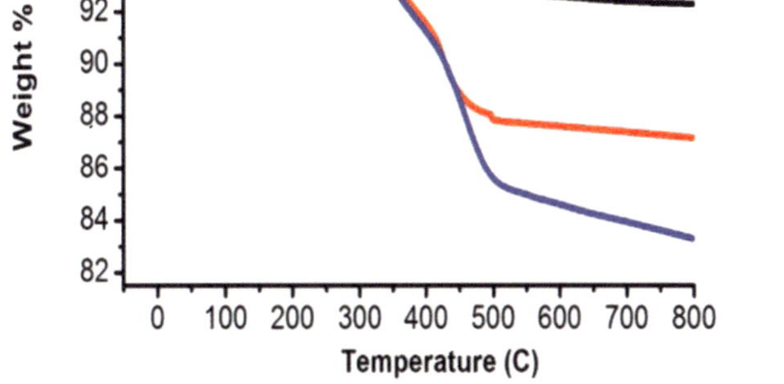

Figure 3.
TGA curves of γ-Al_2O_3-rGO hybrids using calcination time of 1, 2 and 3 h in air atmosphere up to 800°C.

confirmed from the XRD results as shown in **Figures 5** and **6**, respectively. The inset of **Figure 6** shows the XRD of a sample without GO (γ-Al_2O_3). In the XRD spectra of all three samples with calcination times of 1, 2, and 3 h, the presence of characteristic peaks of γ-Al_2O_3 is evident and (matched with JCPDS card no. 10-0425). Further, a relatively broad nature of sharp peaks is observed in the XRD of γ-Al_2O_3 and rGO hybrids (after calcination, **Figure 5**) and in the inset of **Figure 6** (for pure γ-Al_2O_3). This suggests a nanocrystalline structure of γ-Al_2O_3 with nanorod morphology, which is also quite evident from the TEM images (**Figure 4e**).

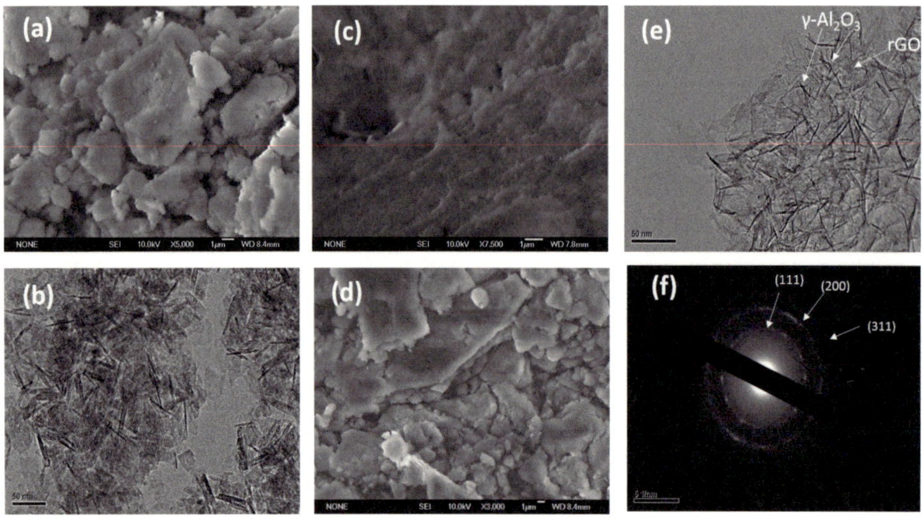

Figure 4.
(a) SEM and (b) TEM images of pure γ-Al_2O_3. SEM images of Al(O)x/GO (c) before calcination at autoclave heating of 453 K for 6 h and (d) after calcination at 723 K for 2 h. (e) TEM image of γ-Al2O3-rGO after calcination at 723 K for 2 h, and (f) SAED pattern of γ-Al_2O_3-rGO hybrid at 723 K for 2 h.

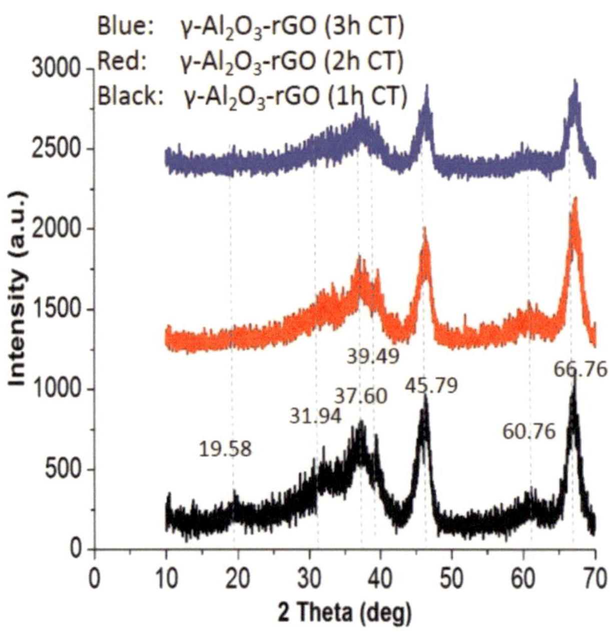

Figure 5.
XRD of γ-Al_2O_3-rGO (1-h calcination time), γ-Al_2O_3-rGO (2-h calcination time) and γ-Al_2O_3-rGO (3-h calcination time).

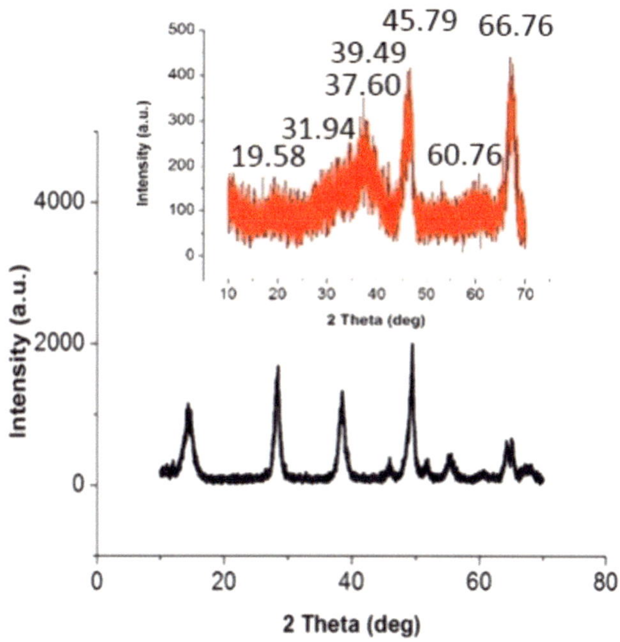

Figure 6.
XRD of γ-Al_2O_3-rGO before calcination and inset is XRD of pure γ-Al_2O_3, fabricated at 723 K.

In all cases after calcination for 1, 2 and 3 h in **Figure 5**, the characteristic peaks of GO that usually appear at 10.28 are also invisible in these hybrids. The total changes in the peaks of the XRD pattern before and after calcination indicate the effects of hydrothermal and calcination treatments.

Raman spectroscopy is conducted on these hybrids to confirm the presence of carbon, shown in **Figure 7**. The Raman spectra of the sample indicate that rGO is present in hybrids with 1-, 2- and 3-h calcination time. For calcinated samples with different times (1, 2, and 3 h), the Raman intensity decreases with calcination time, consistent with TGA results. The G-band value is different in the Raman spectrum of all γ-Al_2O_3 and rGO hybrids as compared to pristine graphene. This reveals the presence of prominent electronic interactions between γ-Al_2O_3 and rGO in hybrids.

The nanohybrids of γ-Al_2O_3-rGO with 3, 2, and 1-h calcination time have BET surface areas of 361, 408, and 379 $m^2•g^{-1}$, respectively. For bare γ-Al_2O_3, the BET surface area is 280 $m^2•g^{-1}$. The bulk densities of γ-Al_2O_3-rGO with 3-, 2-, and 1-h calcination time have values of 1.61, 1.37, and 0.92 g cm^{-3}, respectively. For bare γ-Al_2O_3, the bulk density is 2.75 g cm^{-3}. Clearly, the presence of rGO has led to high BET surface-area values. Presence of rGO in hybrid can increase the BET surface area, pore volume, and thermal conductivity. The higher surface areas, pore volume, thermal and electrical conductivity are significant factors from an applied application point of view. Further as the calcination temperature increased, the crystallinity of γ-Al_2O_3 and rGO hybrids was considerably enhanced as shown in the XRD of γ-Al_2O_3 and rGO hybrids which are taken from 500 to 800 K as in **Figure 8**. For the analysis, calcination time is kept constant (1 h) and further calcination temperature was set as 500, 600, 650, 700, 750, and 800 K, respectively. Actually, most of the samples do not show any characteristics peaks for the rGO in XRD (**Figure 8**).

Now, the crystallinity of γ-alumina is a major concern. Further, it is found that sharp, broad and prominent peaks are obtained for γ-alumina at the higher calcination temperature. But the peaks were weaker or not prominent, when there was a lower calcination temperature such as 500, 600, and 650, respectively.

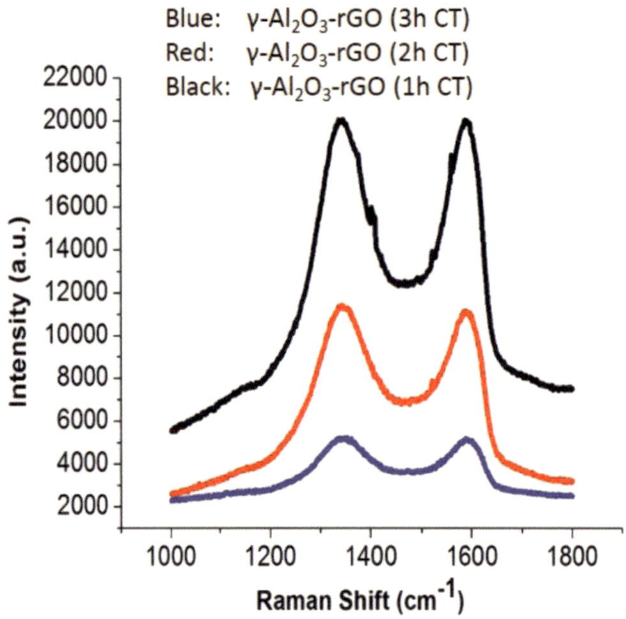

Figure 7.
Raman spectra of γ-Al_2O_3-rGO (1-h calcination time), γ-Al_2O_3-rGO (2-h calcination time) and γ-Al_2O_3-rGO (3-h calcination time).

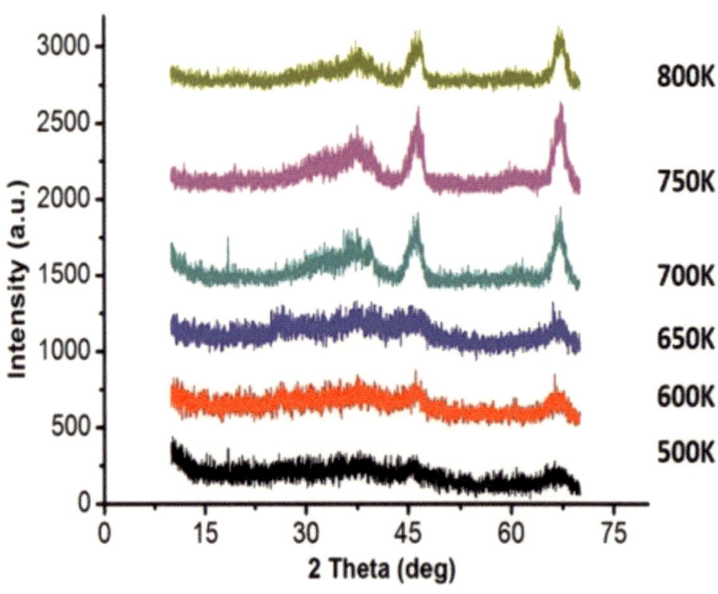

Figure 8.
XRD of γ-Al_2O_3-rGO hybrids taken from 500 to 800 K.

Hot-pressed γ-Al_2O_3 and rGO nanohybrid samples were fabricated at a temperature of 900°C. Hot pressing can affect the quality of graphene. Preserving the quality of graphene as much as possible is a major factor in the enhanced properties of graphene hybrids [19–21]. The SEM morphology of all samples after hot pressing is shown in the **Figure 9**.

The Raman and XRD data for γ-Al_2O_3-rGO calcinated at 2 h, before and after hot pressing, are shown in **Figures 10** and **11**, respectively. Before hot pressing,

Figure 9.
SEM images of hot pressed samples (a) γ-Al_2O_3-rGO (1 h calcination time), (b) γ-Al_2O_3-rGO (2-h calcination time), (c) γ-Al_2O_3-rGO (3-h calcination time), and (d) pure γ-Al_2O_3 (1-h calcination time).

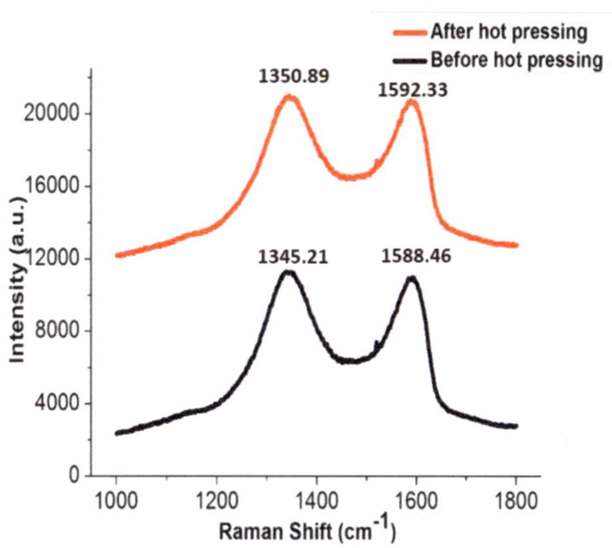

Figure 10.
Raman of γ-Al_2O_3-rGO (2-h calcination time) before and after hot-pressing.

the D band for the sample is found at 1345.21 cm^{-1}. After hot pressing, it is found at 1350.89 cm^{-1}. Before hot pressing, the G band is found at 1588.46 cm^{-1}. After hot pressing, the G band is found at 1592.33 cm^{-1}. This shift in D and G bands is due to the electronic interaction between γ-Al_2O_3 and rGO during hot press processing.

There is not much effect on the integrity of alumina nanorods during hot pressing as this is also confirmed by XRD (**Figure 11**).

The three hot-pressed samples of γ-Al_2O_3-rGO hybrids with calcination times from 1, 2, and 3 h and pure γ-Al_2O_3 (calcinated at 1 h) were studied for properties studies. Electrical conductivity as a function of the concentration of rGO in the hybrid is shown in the **Figure 12**.

It is also found that the electrical conductivity increases with more rGO content. The electrical conductivities of γ-Al_2O_3-rGO calcinated at 1, 2, and 3 h were 8.2×10^1, 7.8×10^1, and 6.7×10^1 $S{\bullet}m^{-1}$, respectively. There is conductivity (5.1×10^{-10} $S{\bullet}m^{-1}$) found in bare γ-Al_2O_3 samples, which confirms that the bare alumina is highly non-conductive. Previous reports show that little carbon (2%) in alumina-carbon hybrids can enhance conductivity up to great level (from 10^{-12} $S{\bullet}m^{-1}$ to 10^{-1} $S{\bullet}m^{-1}$). The improvement of the electrical properties is due to heat treatment, and it has been attributed to mechanisms such as restoration of sp2 C—C bonds and cross-linking between reduced GO sheets during the thermal annealing process [28]. These

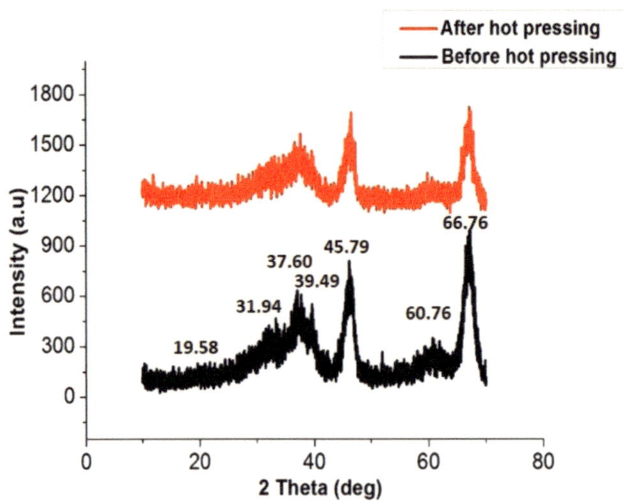

Figure 11.
XRD of γ-Al_2O_3-rGO (2-h calcination time) before and after hot pressing.

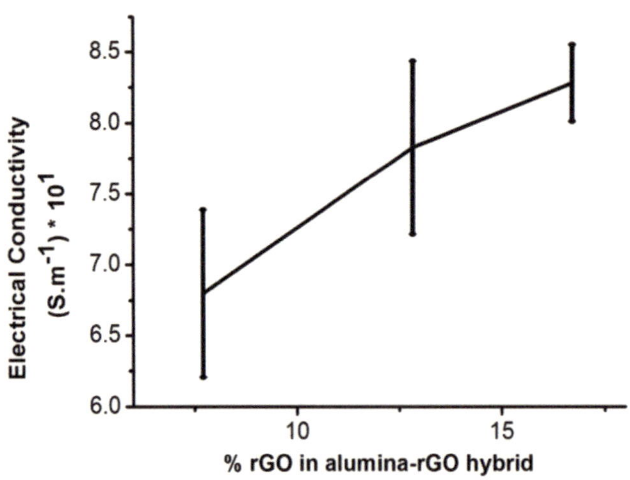

Figure 12.
Electrical conductivity vs. % rGO with error bar.

samples showed a significant increase and improvement in thermal conductivity. This is mainly because of the excess surface electrons and the layered structure of rGO. Thin rGO layers can independently have higher conductivities. The thermal conductivity of pure alumina was reported 0.5 $Wm^{-1} K^{-1}$ at 75°C, which prepared by hydrolysis and solvothermal method. The thermal conductivity of γ-Al_2O_3-rGO with 1-, 2- and 3-h calcination time and bare γ-Al_2O_3 (1-h calcination time) as a function of varying temperatures is shown in **Figure 13**. At a room temperature of 25°C, the thermal conductivities of pure γ-Al_2O_3 and γ-Al_2O_3-rGO (3, 2 and 1-h calcination time) were found to be 0.81, 1.4, 2.37, and 2.53 $Wm^{-1} K^{-1}$, respectively. As the temperature was increased, the thermal conductivity gradually increased in all hybrids of γ-Al_2O_3-rGO and bare γ-Al_2O_3.

In the case of ceramic materials, porosity is one of the main reasons for decreases in the overall thermal conductivity. The dielectric properties of γ-Al_2O_3-rGO hybrids and bare γ-Al_2O_3 were measured using an LCR meter, as shown in **Figure 14**. For γ-Al_2O_3, its dielectric constant is found to be around 9.8, which is closer to that of pure alumina.

For the 3-h calcinated hybrid, the dielectric constant is significantly increased and is multiplied by a factor of 12, which indicates the presence of a first percolation threshold. However, when the rGO content is enhanced in the hybrids by decreasing calcination time to 2 h, the dielectric constant further decreases and approaches the value of pure γ-Al_2O_3. This is attributed to an anomalous trend that produces drastic changes that are usually suffered by most ceramic materials and matrix micro-structures. By further decreasing the calcination temperature of hybrids to 1 h, the dielectric constant increases by four orders of magnitude, which indicates the presence of a second percolation threshold that is achieved through this higher value of dielectric constant. Similarly, the dielectric loss indicates very similar behavior in the real part of the dielectric constant as shown inset of **Figure 14**. Dielectric loss of Al_2O_3-rGO hybrid with 1-h calcination time is much increased, as more rGO in hybrid and this more rGO can make conductive layers' network of rGO in between alumina nanorods; this would cause significant leakage current, and thus result in a high dielectric loss. The existence of a double percolation threshold in such γ-Al_2O_3 and rGO hybrids can be significant for technological and applied applications because it can be used to enhance the dielectric properties in γ-Al_2O_3 and rGO

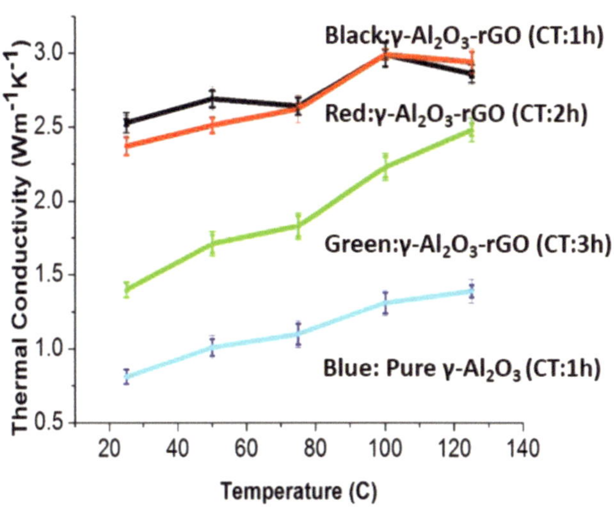

Figure 13.
Thermal conductivity as function of temperature (black, red and green curves are for γ-Al_2O_3-rGO with 1-, 2- and 3-h calcination time with error bar.

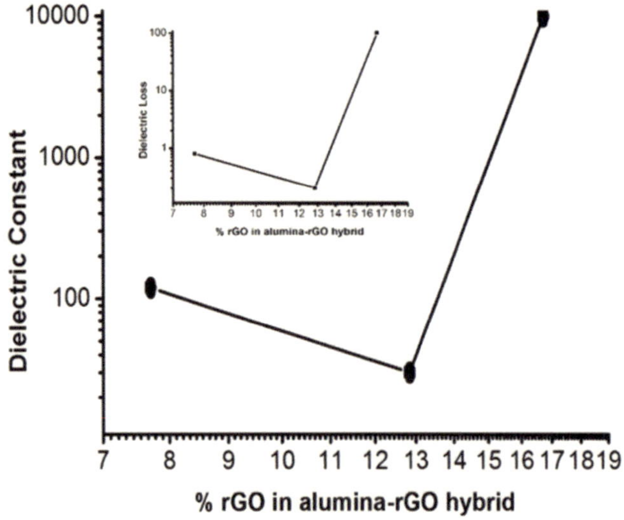

Figure 14.
Dielectric properties vs. % rGO in γ-Al_2O_3-rGO hybrid with error bar.

hybrids with the addition of a small rGO % in the hybrid. From compressive and tensile stress-strain analysis, it is evident that with an increase of rGO content in the hybrid the mechanical compressive and tensile strength is increased as compared to pure alumina. This further caused more strength in alumina hybrids, i.e., higher compressive, tensile strength and higher compressive young modulus values for these hybrids (**Figures 15** and **16**). The enhanced mechanical properties of γ-Al_2O_3 and rGO hybrids can be attributed to covalent interaction of rGO with γ-Al_2O_3 and to efficient load transfers between rGO and nanorods of γ-Al_2O_3. Further, this is closely bound with the elongated and fine γ-Al_2O_3 nanorods and atomic-level rGO layers with a covalent interaction with γ-Al_2O_3. Young modulus of γ-Al_2O_3-rGO with 1-, 2- and 3-h calcination time and γ-Al_2O_3 with 1-h calcination time are calculated as 3.7, 3.2, 2.65 and 1.80 GPa. In this case, lower tensile and compressive strength in alumina can be due to the availability of powder instead of single crystals of alumina. Increase in calcination temperature has reduced wt.% of rGO in a hybrid. This is the reason of having more strength in hybrids with lower calcination

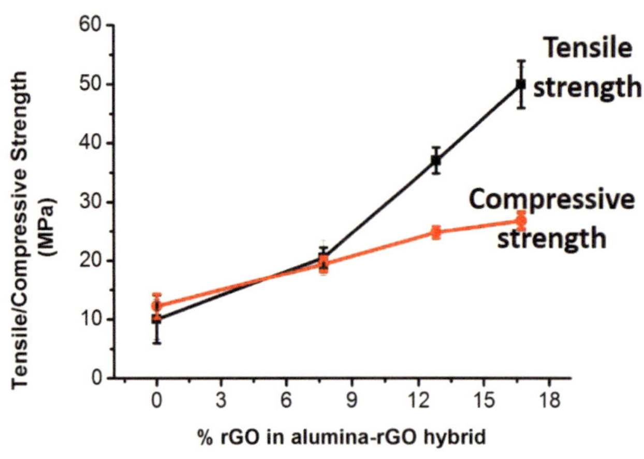

Figure 15.
Compressive and tensile strength as a function of rGO for hybrid γ-Al_2O_3-rGO with error bars.

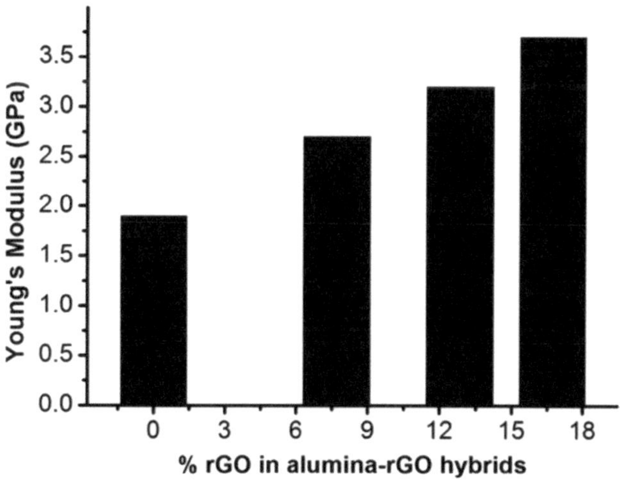

Figure 16.
Compressive Young's modulus as % rGO in alumina-rGO hybrids.

temperature. The maximum value of Young's modulus (3.7 GPa) is determined in 1-h calcinated alumina-rGO hybrid, as also shown in **Figure 16**.

Thus, elongated dimensions of nanorods are a major cause of higher mechanical strength in these hybrids. In γ-Al_2O_3-rGO monoliths, higher calcination temperature enhances length, diameter and aspect ratios of γ-Al_2O_3 nanorods. Presence of more rGO and higher aspect ratio elongated alumina rods determines the interface interaction between rGO platelets and alumina. A 90% increase in tensile strength and 75% in compressive strength occurs when the content of rGO is increased from 0 to 7.705 wt.% in the hybrid calcination of a hybrid at 3-h processing time. With the increase of rGO, alumina-rGO hybrids have shown higher values for young modulus. The hybrids with 1-h calcination time show good enhancement in its electrical conductivity (8.2×10^1 S m^{-1}) due to the availability of more surface electrons of rGO. This is best-reported values for conductivity. After hot press process, there is a wide increase in electrical conductivity values when there is a decrease of calcination temperature from 3 to 1-h processing time in these hybrids. Further, the thermal conductivity of γ-Al_2O_3-rGO is enhanced by more than 80% compared to that of bare γ-Al_2O_3 when there is an increase in rGO content up to 7.705 wt.% in γ-Al_2O_3 and rGO hybrids. There is a 77% increase in thermal conductivity using this solvothermal method in these γ-Al_2O_3 and rGO hybrids. Physical properties such as the BET surface area and bulk density are also improved. Elongated dimensions of nanorods are a major cause of higher mechanical strength in these hybrids. Dielectric constant increases by four orders of magnitude through second percolation threshold with the addition of small rGO in a hybrid. Enhancement in physical properties can be due to well aligned, elongated and fine nanorods morphology of alumina in hybrids, calcination, and hot press processing further played an important role by sustaining quality rGO in hybrids. These nanohybrids of alumina monoliths and rGO can be further applied as catalysts, electrolytes, and as electrochemically active materials because of their nanometer dimensions and enhanced physical properties.

4. Physical properties of silica-graphene hybrids for technological applied applications

With rGO in the SiO_2-rGO hybrids, the hybrids powder shows a prominent change in the color after calcination. TGA of SiO_2-rGO powder samples shows that different initial GO suspensions (0.1, 0.2, and 0.3 g) have led to different concentrations of rGO in the hybrids. The TGA curves of all hybrids show a stable weight loss up to 100°C, as a result of moisture loss and between 150 and 300°C, as a removal of unreduced GO functional groups, and from 350 to 600°C, as a result of removal of all carbon related materials which are due to the decomposition of rGO, and other impurities (if any) after heating SiO_2-rGO hybrids to 800°C in an air atmosphere. By keeping 1-h calcination time, with initial 0.1 g GO suspension in hybrid, the wt.% loss of 1.55 was calculated. With initial 0.2 g GO suspension in hybrid, the wt.% loss of 6.75 was calculated. With initial 0.3 g GO suspension in hybrid, the wt.% loss of 10.82 was calculated. From the wt.% loss, it is found that % rGO is determined as 1.55, 6.75 and 10.82 in the SiO_2-rGO hybrids, which are obtained from 0.1, 0.2 and 0.3 g GO suspensions, respectively. All powder samples before calcination (at 800 K for 1 h) but after autoclave heating (at 420 K for 4 h) have been referred to as Si(O)x/rGO. Lower and higher magnification SEM images for hybrid SiO_2-rGO-1.55% (calcinated at 800 K for 1 h) are shown as in **Figure 17a** and **b**. From SEM images, it is clear that SiO_2 have sphere morphology with location side by side along with overlapping and wide size distribution of spheres, ensuring very close contact among SiO_2 spheres in the whole network. The diameters of spheres have variation in size ranging from 5 nm to 3 μm. Bare SiO_2 particles (calcinated at 800 K for 1 h) have shown spherical morphology with the little rough surface as shown in the **Figure 17b** and **d**. The rough texture of SiO_2 sphere is confirmed by the closely looking to **Figure 17d**. Diameters of SiO_2 spheres vary in size from 1 μm to a few micrometers. TEM is carried out to verify the morphology, which is obtained in SEM. Further, TEM image of SiO_2-rGO-1.55% shows that the SiO_2 spheres were uniformly encapsulated by the thin layers of rGO (marked with an arrow) as in the **Figure 17e**. After hydrothermal (420 K for 4 h) and calcination (800 K for 1 h) treatments, in SiO_2-rGO-1.55% hybrid, the SiO_2 spheres is fully decorated with small rGO sheets.

In **Figure 17e**, the low-contrast features are small portions of the graphene layers on which SiO_2 spheres are uniformly distributed. After thermal treatments, uniform distribution of SiO_2 spheres and rGO sheets is the main factor for enhanced properties of the SiO_2-rGO hybrid. TEM image of SiO_2-rGO-1.55% is shown in the

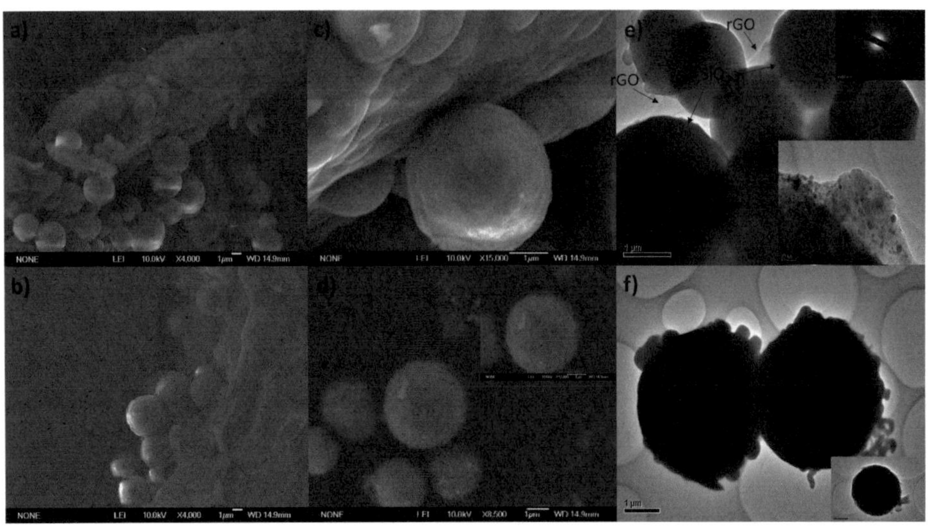

Figure 17.
SEM images of (a) SiO_2-rGO-1.55% (b) pure SiO_2 fabricated at a calcination temperature of 800 K for 1 h at lower magnification. SEM images of same (c) SiO_2-rGO-1.55% (d) pure SiO_2 at higher magnification. (e) TEM images of same SiO_2-rGO-1.55% at lower magnifications, inset is SAED pattern for same sample and (f) TEM images of same pure SiO_2 at lower magnifications.

Figure 17e, from TEM image it is evident that diameters of SiO_2 spheres ranges from few nanometers to few micrometers, and agreed with SEM results. For the SiO_2-rGO-1.55% hybrid, the selected area electron diffraction (SAED) pattern is presented in the inset of **Figure 17e**, and it shows the hybrid is composed of SiO_2 spheres and thin rGO layers. In SAED pattern, the electron diffraction rings indicate the amorphous nature of SiO_2. Rough texture and the non-uniform surface of a pure SiO_2 sphere are confirmed from TEM results in **Figure 17f**. Diameters of spherical SiO_2 vary from 1 μm to a few micrometers as evident from TEM characterization (**Figure 17f**, and in agreement with SEM results of **Figure 17b** and **d**. For more insight into structural properties of SiO_2-rGO hybrids, the fabrication of the amorphous SiO_2 was confirmed from the XRD results as shown in the **Figure 18**. XRD peak of the pure SiO_2 confirmed the presence of amorphous SiO_2 in the sample, which is prepared from the same method and under same experimental conditions but without the addition of GO.

In the XRD spectra of all the samples with % rGO of 1.55, 6.75 and 10.82 and pure SiO_2, the presence of characteristic peaks of amorphous SiO_2 is evident and matched with JCPDS card no. 01-086-1561. Relatively intense and broader peaks are observed in the XRD of pure SiO_2, SiO_2-rGO-1.55% and SiO_2-rGO-6.75% hybrids in comparison to SiO_2-rGO-10.82% hybrid. In XRD pattern (**Figure 18**), the absence of XRD peak at 10.28o is the evidence of the successful reduction of GO in hybrids after hydrothermal and calcination treatments. Raman spectroscopy confirmed the presence of carbon in hybrids. Meanwhile, SiO_2-rGO hybrids were tested for FTIR spectra to evaluate the reduction of GO and successful attachments of SiO_2 spheres with rGO platelets, as shown in **Figure 19**. With 1.55, 6.75 and 10.82 wt.% rGO, it is found that SiO_2-rGO hybrids have shown IR band at 3417 and 1396 cm^{-1}, which are mainly due to the stretching and bending vibrations of —OH structural group and stretching vibrations of absorbed water.

All hybrids have shown the IR band at 1642 cm^{-1}, which is due to the bending vibration of absorbed water molecules. In SiO_2-rGO hybrids, the IR bands at 1110, 800 and 479 cm^{-1} are assigned to the Si-O-Si asymmetric stretching and bending vibrations, respectively. According to previous reports, the IR bands at 1110, 800 and 479 cm^{-1} are main IR characteristics bands for amorphous nature of silica, and such bands appear in IR spectra of all three SiO_2-rGO hybrids. In hybrids, IR band at 380 cm^{-1} is associated with the Si-OH wagging mode. In-between 500 and 780 cm^{-1}, the appearance of IR bands is mainly because of assembling of SiO_2 spheres and rGO platelets in hybrids. Below 1000 cm^{-1}, IR bands are mainly due

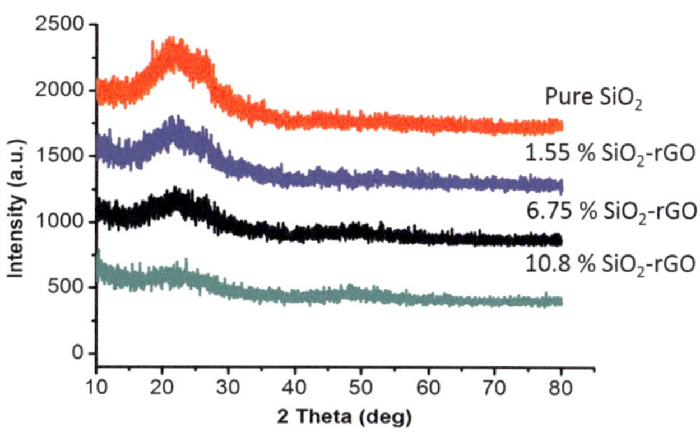

Figure 18.
XRD of SiO_2-rGO hybrids with 1.55, 6.75 and 10.8% rGO and pure SiO_2 fabricated at a calcination temperature of 800 K for 1 h.

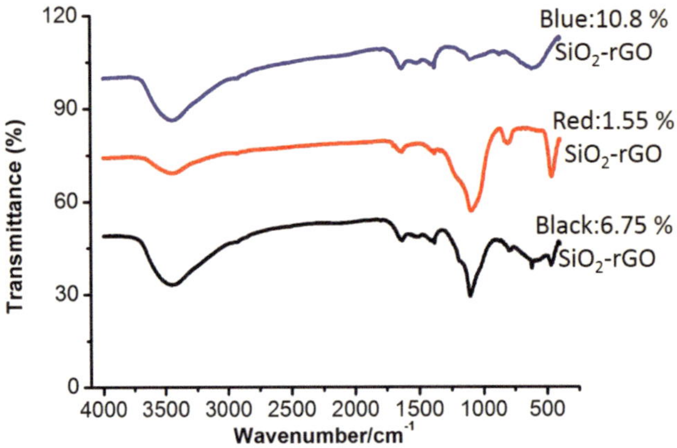

Figure 19.
FTIR of SiO_2-rGO hybrids with 1.55, 6.75 and 10.8% rGO.

to reduced oxygen and —OH groups, which is a further indication of a successful reduction of GO in hybrids after hydrothermal and calcination treatments. FTIR spectra of all hybrids possess less pronounced IR bands, and which are associated with the presence of amorphous silica. Amorphous silica can be distinguished from crystalline silica based on far FTIR region measurements. For more information of the chemical composition and elemental states, the XPS spectroscopy of hybrids is carried out. With 1.55, 6.75 and 10.82 wt.% rGO, XPS spectra of SiO_2-rGO hybrids is presented in **Figure 20**.

In case of all the hybrids, XPS spectra show distinct peaks for C1s, O1s, O2s, Si2s, and Si2p. In case of SiO_2-rGO-10.82% hybrid, the intensity ratios of the C1s peak to the Si 2 s and Si 2p peaks are measured, and intensity ratios are higher in comparison to hybrids with 6.75 and 1.55% rGO. This may be due to more rGO in hybrids, as confirmed by the TGA results. The samples were tested for Brunauer-Emmett-Teller (BET) surface areas. In hybrids, higher % of rGO helps to increase surface area compared to that of pure SiO_2. With 1.55, 6.75 and 10.82% rGO, the nanohybrids of SiO_2-rGO have BET surface areas of 611.21 ± 19.02, 677.53 ± 25.21, and 712.01 ± 13.21 $m^2 \bullet g^{-1}$, respectively. For pure SiO_2, the BET surface area is 333.07 ± 21.57 $m^2 \bullet g^{-1}$. With 1.55, 6.75 and 10.82% rGO, the bulk densities of SiO_2-rGO have values of 1.41 ± 0.07, 1.29 ± 0.19, and 0.89 ± 0.03 g cm^{-3}, respectively. For bare SiO_2, the bulk density is 2.75 ± 0.12 g cm^{-3}. It is found that mesoporosity

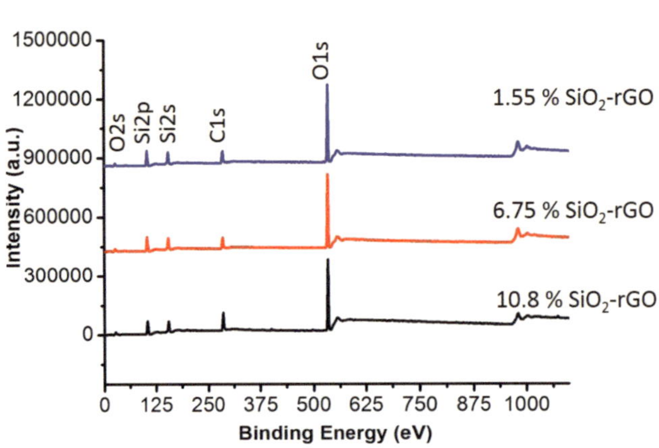

Figure 20.
XPS spectra of SiO_2-rGO hybrids with 1.55, 6.75 and 10.8% rGO.

is highest with mesopores volume % of 90.52, which is measured for pure SiO_2. It is quite clear that minimum mesopores volume % is 57.11, which is calculated for SiO_2-rGO hybrid with 10.82% rGO. With higher wt.% of SiO_2, hybrids have shown an increase of mesopores volume %. Clearly, the presence of more rGO has led to higher BET surface area in hybrids. Well-aligned SiO_2 spheres with ultrathin rGO platelets resulted in a higher surface area and lower density compared to pure SiO_2. In hybrids, the presence of more rGO has increased the BET surface area, which is the main reason for enhanced thermal and electrical conductivity due to the availability of more surface electrons of rGO platelets. The higher value of surface areas, electrical conductivity and thermal conductivity are significant factors for applications in thermal and electrical engineering. The effect of hot pressing pressure on SiO_2 and rGO hybrid was conducted. Raman spectroscopy is conducted on SiO_2-rGO-6.75, which are hot pressed at 10, 20 and 30 MPa, respectively. The higher hot pressing pressure led the D peak intensity to reduce. For SiO_2-rGO-6.75 hybrid at 30 MPa, D peak is almost disappeared as compared to the hot-pressing pressure of 10 and 20 MPa. It shows that higher hot pressing pressure leads to remove defects from the rGO platelets and reconstruction of sp2 carbon network in hybrid. The D/G intensity ratio has reduced with higher hot pressing pressure. Solvothermal-hot press processed SiO_2 and rGO hybrids have shown a significant improvement in the physical properties such as the electrical conductivity, thermal conductivity, tensile strength and the dielectric constant. In SiO_2-rGO-10.8% hybrid, the electrical conductivity of 0.143 $S{\bullet}m^{-1}$ is calculated. For the same sample, the thermal conductivity of 1.612 $Wm^{-1}\,K^{-1}$ is calculated, which is enhanced by more than 255%. With little rGO in hybrid, dielectric constant increases by seven orders of magnitude through second percolation threshold, such values for dielectric constants are higher. In solvothermal-hot press processed SiO_2-rGO, improved physical properties are due to more rGO platelets, high-temperature calcination, high-temperature hot pressing and formation of conductive pathways between rGO platelets and SiO_2 spheres.

5. Conclusions

This book chapter has explained ceramics-graphene hybrids, enhanced properties and possible applications in ceramics-graphene industry. Further, by the solvothermal-hot pressing method, a complete systematic study on enhanced physical properties of the hybrids has been made, which can further implement hybrids in advanced technological applications. This can lead to a significant contribution for the applications of ceramics-graphene assembly nanomaterials which can be further applied as electrolytes, catalysts, conductive, electrochemically active, and as dielectric materials for the high-temperature applications due to enhanced physical properties.

Conflict of interest

Authors have declared no "conflict of interest."

Author details

Mujtaba Ikram[1*] and Muhammad Umer Farooq[2]

1 Department of Physics, Government College University (GCU), Lahore, Pakistan

2 Department of Physics, University of Education, Faisalabad Campus, Faisalabad, Pakistan

*Address all correspondence to: dr.mujtabaikram@gcu.edu.pk

References

[1] Sau TK, Rogach AL, Jäckel F, Klar TA, Feldmann J. Properties and applications of colloidal nonspherical noble metal nanoparticles. Advanced Materials. 2010;**22**:1805-1825. DOI: 10.1002/adma.200902557

[2] Xia Y, Xiong Y, Lim B, Skrabalak SE. Shape-controlled synthesis of metal nanocrystals: simple chemistry meets complex physics. Angewandte Chemie International Edition. 2009;**48**:60-103. DOI: 10.1002/anie.200802248

[3] Liu ZB, Xu YF, Zhang XY, Zhang XL, Chen YS, Tian JG. Porphyrin and fullerene covalently functionalized graphene hybrid materials with large nonlinear optical properties. The Journal of Physical Chemistry B. 2009;**113**:9681-9686. DOI: 10.1021/jp9004357

[4] Sau TK, Rogach AL. Nonspherical noble metal nanoparticles: colloid-chemical synthesis and morphology control. Advanced Materials. 2010;**22**:1781-1804. DOI: 10.1002/adma.200901271

[5] Markandan K, Chin JK, Tan MT. Recent progress in graphene based ceramic composites: a review. Journal of Materials Research. 2017;**32**:84-106. DOI: 10.1557/jmr.2016.390

[6] Neto AC, Guinea F, Peres NM, Novoselov KS, Geim AK. The electronic properties of graphene. Reviews of Modern Physics. 2009;**81**:109. DOI: 10.1103/RevModPhys.81.109

[7] Kim KS, Zhao Y, Jang H, Lee SY, Kim JM, Kim KS, et al. Large-scale pattern growth of graphene films for stretchable transparent electrodes. Nature. 2009;**457**:706-710. DOI: 10.1038/nature07719

[8] Novoselov KS, Fal V, Colombo L, Gellert P, Schwab M, Kim K. A roadmap for graphene. Nature. 2012;**490**:192-200. DOI: 10.1038/nature11458

[9] Novoselov KS, Geim AK, Morozov SV, Jiang D, Zhang Y, Dubonos SV, et al. Electric field effect in atomically thin carbon films. Science. 2004;**306**:666-669. DOI: 10.1126/science.1102896

[10] Benavente R, Pruna A, Borrell A, Salvador M, Pullini D, Peñaranda-Foix F, et al. Fast route to obtain Al2O3-based nanocomposites employing graphene oxide: synthesis and sintering. Materials Research Bulletin. 2015;**64**:245-251. DOI: 10.1016/j.materresbull.2014.12.075

[11] Bhowmik K, Chakravarty A, Bysakh S, De G. γ-Alumina nanorod/reduced graphene oxide as support for poly(ethylenimine) to capture carbon dioxide from flue gas. Energy Technology. 2016;**4**:1409-1419. DOI: 10.1002/ente.201600186

[12] Chen L, Chai S, Liu K, Ning N, Gao J, Liu Q, et al. Enhanced epoxy/silica composites mechanical properties by introducing graphene oxide to the interface. ACS Applied Materials & Interfaces. 2012;**4**:4398-4404. DOI: 10.1021/am3010576

[13] Huang D, Yang Z, Li X, Zhang L, Hu J, Su Y, et al. Three-dimensional conductive networks based on stacked SiO2@graphene frameworks for enhanced gas sensing. Nanoscale. 2017;**9**:109-118. DOI: 10.1039/C6NR06465E

[14] Kumar KS, Sanyadanam S, Paik P. Dangling ultrafine nano silica on graphene oxide to form hybrid nanocomposite: enhancement of dielectric properties. Materials Research Express. 2016;**3**:055019. DOI: 10.1088/2053-1591/3/5/055019

[15] Lv L, Huang L, Zhu P, Li G, Zhao T, Long J, et al. SiO2 particle-supported

ultrathin graphene hybrids/polyvinylidene fluoride composites with excellent dielectric performance and energy storage density. Journal of Materials Science: Materials in Electronics. 2017;**28**:13521-13531. DOI: 10.1007/s10854-017-7191-0

[16] Watcharotone S, Dikin DA, Stankovich S, Piner R, Jung I, Dommett GH, et al. Graphene-silica composite thin films as transparent conductors. nano letters. 2007;**7**:1888-1892. DOI: 10.1021/nl070477

[17] Qian R, Yu J, Wu C, Zhai X, Jiang P. Alumina-coated graphene sheet hybrids for electrically insulating polymer composites with high thermal conductivity. RSC Advances. 2013;**3**:17373-17379. DOI: 10.1039/c3ra42104j

[18] Walker LS, Marotto VR, Rafiee MA, Koratkar N, Corral EL. Toughening in graphene ceramic composites. ACS Nano. 2011;**5**:3182-3190. DOI: 0.1021/nn200319d

[19] Akhtar F, Rehman Y, Bergström L. A study of the sintering of diatomaceous earth to produce porous ceramic monolithswith bimodal porosity and high strength. Powder Technology. 2010;**201**:253-257. DOI: 10.1016/j.powtec.2010.04.004

[20] Bhowmik K, Pramanik S, Medda SK, De G. Covalently functionalized reduced graphene oxide by organically modified silica: a facile synthesis of electrically conducting black coatings on glass. Journal of Materials Chemistry. 2012;**22**:24690-24697. DOI: 10.1039/c2jm35429b

[21] Dervin S, Lang Y, Perova T, Hinder SH, Pillai SC. Graphene oxide reinforced high surface area silica aerogels. Journal of Non-Crystalline Solids. 2017;**465**:31-38. DOI: 10.1016/j.jnoncrysol.2017.03.030

[22] Porwal H, Grasso S, Reece M. Review of graphene–ceramic matrix composites. Advances in Applied Ceramics. 2013;**112**:443-454. DOI: 10.1179/174367613X13764308970581

[23] Lwin S, Li Y, Frenkel AI, Wachs IE. Nature of WO*x* sites on SiO2 and their molecular structure–reactivity/selectivity relationships for propylene metathesis. ACS Catalysis. 2016;**6**:3061-3071. DOI: 10.1021/acscatal.6b00389

[24] Merle N, Le-Quéméner FDR, Bouhoute Y, Szeto KC, De-Mallmann A, Barman S, et al. Well-defined molybdenum oxo alkyl complex supported on silica by surface organometallic chemistry: a highly active olefin metathesis precatalyst. Journal of the American Chemical Society. 2017;**139**:2144-2147. DOI: 10.1021/jacs.6b11220

[25] Ciriminna R, Fidalgo A, Pandarus V, Béland F, Ilharco LM, Pagliaro M. The sol–gel route to advanced silica-based materials and recent applications. Chemical Reviews. 2013;**113**:6592-6620. DOI: 10.1021/cr300399c

[26] Huang L, Zhu P, Li G, Lu DD, Sun R, Wong C. Core–shell SiO2@RGO hybrids for epoxy composites with low percolation threshold and enhanced thermo-mechanical properties. Journal of Materials Chemistry A. 2014;**2**:18246-18255. DOI: 10.1039/C4TA03702B

[27] Ramezanzadeh B, Haeri Z, Ramezanzadeh M. A facile route of making silica nanoparticles-covered graphene oxide nanohybrids (SiO2-GO); fabrication of SiO2-GO/epoxy composite coating with superior barrier and corrosion protection performance. Chemical Engineering Journal. 2016;**303**:511-528. DOI: 10.1016/j.cej.2016.06.028

[28] Huang D, Li X, Wang S, He G, Jiang W, Hu J, et al. Three-dimensional chemically reduced graphene oxide templated by silica spheres for ammonia sensing. Sensors and Actuators B:

Chemical. 2017;**252**:956-964. DOI: 10.1016/j.snb.2017.05.117

[29] Zhibin L, Xue-Feng Y, Zhu P-K. Recent advances in cell-mediated nanomaterial delivery systems for photothermal therapy. Journal of Materials Chemistry B. 2018;**6**:1296-1311. DOI: 10.1039/C7TB03166A

[30] Islam A, Mukherjee B, Sribalaji M, Rahman OA, Arunkumar P, Babu KS, et al. Role of hybrid reinforcement of carbon nanotubes and graphene nanoplatelets on the electrical conductivity of plasma sprayed alumina coating. Ceramics International. 2017;**44**:4508-4511. DOI: 10.1016/j.ceramint.2017.12.021